ARMY ARCHITECTURE IN THE WEST

Published in association with the Center for American Places,
Santa Fe, New Mexico, and Harrisonburg, Virginia

Army Architecture in the West

Forts Laramie, Bridger, and
D. A. Russell, 1849–1912

Alison K. Hoagland
Foreword by Paul L. Hedren

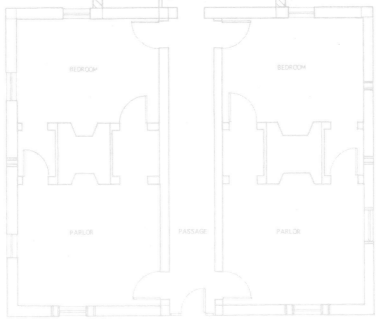

UNIVERSITY OF OKLAHOMA PRESS : NORMAN

This book is published with the generous assistance of The McCasland Foundation, Duncan, Oklahoma.

A version of chapter 3 appeared in *Winterthur Portfolio* 34.4 (winter 1999): 215–37. A version of chapter 4 appeared in the *Journal of the Society of Architectural Historians* 57.3 (September 1998): 298–315.

Library of Congress Cataloging-in-Publication Data

Hoagland, Alison K., 1951–
    Army architecture in the West : Forts Laramie, Bridger, and D. A. Russell, 1849–1912 / Alison K. Hoagland ; foreword by Paul L. Hedren.
      p. cm.
    Includes bibliographical references and index.
    ISBN 0-8061-3620-0 (alk. paper)
    1. Fortification—West (U.S.) 2. Military architecture—West (U.S.) 3. Fort Laramie (Wyo.) 4. Fort Bridger (Wyo.) 5. Fort D. A. Russell (Wyo.) I. Title.

UG411.W48H63 2004
725'.18—dc22

2004045988

The paper in this book meets the guidelines for permanence and durability of the Committee on Production Guidelines for Book Longevity of the Council on Library Resources, Inc. ∞

Copyright © 2004 by the University of Oklahoma Press, Norman, Publishing Division of the University. All rights reserved. Manufactured in the U.S.A.

1 2 3 4 5 6 7 8 9 10

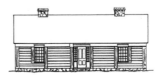

CONTENTS

Acknowledgments	vii
Foreword, by Paul L. Hedren	ix
Introduction	3

Part I. Outpost: Forts Laramie and Bridger, 1849–1869

1. An Oasis in the Desert: Purpose and Plan	13
2. Huts and Adobe Buildings Sadly in Need of Repair: Architecture	39

Part II. Village: Forts Laramie, Bridger, and D. A. Russell, 1867–1890

3. A Beautiful Village: Purpose and Plan	79
4. The Homelike Appearance of the House: Architecture	119
5. Quite an Air of Home Comfort: Furnishings and Conveniences	172

Part III. Institution: Fort D. A. Russell, 1890–1912

6. A Monument to the Pork Barrel: Purpose, Plan, and Architecture	203

Epilogue	244
Notes	251
Bibliography	269
Index	283

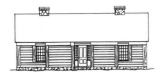

ACKNOWLEDGMENTS

My first and largest debt of gratitude is to Thomas Carter, who first suggested that I write this book. Tom helped shape this project, and his comments and suggestions have always been insightful and thought-provoking. His enthusiasm for this project and his wise counsel have been unending. Tom also devoted a summer field school of the University of Utah's Graduate School of Architecture to documenting forts with me, work supported in part by the National Endowment of the Humanities. Thanks to those students: Jim Agutter, Liza Hart, Barrett Powley, and Heather Randall of the University of Utah, and Kevin O'Dell from the graduate program in industrial archaeology at Michigan Technological University.

At the forts, I received immeasurable assistance from Sandra Lowry, as well as other staff, at Fort Laramie National Historic Site; Linda Byers and Dudley Gardner at Fort Bridger State Historic Site; Jeff Hauff of the State Parks and Historic Sites Division; and Richard Bryant at F. E. Warren Air Force Base. Staff at the Wyoming State Archives, American Heritage Center, Library of Congress, and National Archives were all extremely helpful. For lodging on research trips, I would like to thank Mark Schara and Gray Fitzsimons in Washington, D.C., and Carole and Tom Eppler of the Porch Swing Bed and Breakfast in Cheyenne. For assists at crucial points, thanks also to Allen Chambers, Gordon Chappell, Tim Davis, Dale Floyd, Richard L. Hayes, Sarah Heald, Dean Herrin, Kate Kuranda, Sara Amy Leach, Bob Rea, Pamela Scott, and Chris Wilson.

Michigan Technological University was my home base during the research and writing. I thank my colleagues there, particularly Terry Reynolds, Mary Durfee, and Carol Maclennan, for their support and interest. Larry Lankton deserves a special mention for his good advice, constant mentoring, and astute reading of my work. At the Van Pelt Library, my thanks go to Barbara Wilder and Stephanie Pepin of the Inter-Library Loan division, and Sherry Engle of Government Documents, who helped me overcome the disadvantages of a remote location.

Finally, I acknowledge the support and influence of my late mother, Cynthia Kimball Hoagland. On my spring breaks during high school and college, she introduced me to that venerable institution, the Road Trip, and we headed out to look at historic buildings up and down the East Coast. I am sure that my interest in architectural history is due to those tours. Several years ago, while I was researching forts on the northern Great Plains that had similarities to Fort D. A. Russell, I returned the favor by inviting her along on a road trip through Minnesota, North Dakota, eastern Montana, and northern Wyoming, before joining my father for some fishing near Helena, Montana. We had great fun. I dedicate this book to her memory.

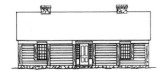

FOREWORD

The forts of the American West have fascinated people ever since these imposing edifices were eclipsed from the national scene. Western posts were distinctive, emblematic, and even mystical. Some, such as the American Fur Company's Fort Union in North Dakota or the U.S. Army's Fort Snelling in Minnesota, conjure images of empire; others, such as Fort Kearney, Nebraska, and early Fort Laramie, Wyoming, served as havens on the long, dusty roads west; still others, such as Fort Phil Kearny, Wyoming, and early Camp Robinson, Nebraska, functioned as citadels in a strife-torn land or, like Forts D. A. Russell and Mackenzie in Wyoming, quartered soldiers in stately garrison "resorts." Artists and film and television producers would have us believe that peril lurked just beyond the sturdy gates of every one of those universally stockaded outposts. But Hollywood has never troubled to differentiate between trading and military forts, and has never acknowledged that few army forts were walled, and fewer yet were ever attacked by Indians.

Dispelling these myths is regular business at many of today's historic forts, a remarkable number of which survive as parks managed by federal, state, and local agencies. But the answers are far from standard, because the truths about these structures are wonderfully complex. The army's Fort Laramie never had a wall, and modern visitors are sadly disappointed at not finding one. Fort Phil Kearny on the bloody Bozeman Trail had a stockade but was never attacked by Indians. The denizens of Fort Ridgely, Minnesota, were not protected by a wall, but in 1862 were twice viciously attacked by Dakota tribesmen. Fort Union had a wall, but its gates were closed nightly only to corral the livestock, and were opened by day to gladly welcome in Assiniboine, Cree, and Crow Indians to trade.

Providing accurate information to visitors is all the more challenging because the fort builders themselves followed no standard guidelines as they undertook their business. They had to choose sites out of immediate need and hope that the location would provide such necessities as wood and water. They used no conventional structural plans, but rather bowed to whim, available labor, and meager budgets for such expensive items as imported brick, glass, and hardware. Army forts invariably had a discernible form: four sides on a central green, with officer quarters lining one or two sides, enlisted barracks another, and utilitarian structures like the hospital, storehouses, and guardhouse on the fourth, all surrounding an open rectangular parade ground. Topping all, usually from a corner or center of the parade ground, was a towering flagpole and the Stars and Stripes, uniquely marking a garrison home of the U.S. Army.

A wonderful array of reminiscences and histories of life at western forts and of the posts themselves has emerged since the days of abandonment. Elizabeth Custer and Elizabeth Burt are among many officers' wives regaling us with stories of the hardships, gaiety, and anxiety known by inhabitants. A famed Indian war campaigner turned novelist, Charles King, penned several compelling histories and dozens of novels featuring aspects of life at frontier army posts, including Forts Laramie and D. A. Russell on the northern plains. And modern-day historians are methodically adding to an ever growing bibliography of works on western forts. Forts Snelling, Riley, Leavenworth, and Laramie are among those whose histories were written long since by prominent scholars, but now such locations as Fort Union, North Dakota, Fort Fetterman, Wyoming, and Fort Robinson, Nebraska, are chronicled in more recent works by Bart Barbour, Tom Lindmier, and Tom Buecker, respectively. Other newer works include histories of Fort C. F. Smith, Montana, by Barry Hagen; of Forts Caspar and Mackenzie, Wyoming, by John McDermott; and a forthcoming history of Fort Randall, South Dakota, by Jerome Greene.* To be

---

* Barton H. Barbour, *Fort Union and the Upper Missouri Fur Trade* (Norman: University of Oklahoma Press, 2001); Tom Lindmier, *Drybone: A History of Fort Fetterman, Wyoming* (Glendo, Wyo.: High Plains Press, 2002); Thomas R. Buecker, *Fort Robinson and the American West, 1874–1899* (Lincoln: Nebraska State Historical Society, 1999); Buecker, *Fort Robinson and the American Century, 1900–1948* (Lincoln: Nebraska State Historical Society, 2002); Barry J. Hagan, *Exactly in the Right Place: A History of Fort C. F. Smith, Montana Territory, 1866–1868* (El Segundo, Calif.: Upton & Sons, Publishers, 1999); John D. McDermott, *Frontier Crossroads: The History of Fort Caspar and the Upper Platte Crossing* (Casper: City of Casper, Wyoming, 1997); McDermott, *Fort Mackenzie: A Century of Service, 1898–1998* (Sheridan, Wyo.: Fort Mackenzie Centennial Committee, 1998).

sure, there are still tales to be written and rewritten about prominent posts like Forts Abraham Lincoln, Assinniboine, Keogh, and McKinney, and even little outposts like Fort Hartsuff, Camp on Sage Creek, Glendive Cantonment, and Bad Lands Cantonment. These and many other northern plains army outposts deserve their own chroniclers.

Until now, we have also wished for a better understanding of the manner in which these posts were constructed. They followed form, but exactly what form? For decades forts featured eclectic internal architecture, but why? While the army in the most general sense built its own garrison homes—literally, with soldiers cutting the lumber, wielding hammers and hods, and whitewashing the finishes—someone was in charge who bid for and appropriated the funding, determined spatial allocations, laid out the posts and each interior construction, and supervised the skilled and unskilled laborers. And someone else eventually standardized the army's plans, bargained for centralization, reordered the manner in which forts were constructed across the land, and ultimately engaged in the political wrangling always at work in military construction. (The interdependence of local communities and neighboring army garrisons is a timeless story in America, with a particularly powerful twist at Fort D. A. Russell at the turn of the twentieth century.)

Now we have Alison Hoagland's *Army Architecture in the West: Forts Laramie, Bridger, and D. A. Russell, 1849–1912*, a distinctive, definitive study of fort building in Wyoming. Her work studies three legendary posts, each nationally significant in its own right, and collectively emblematic of self-directed army industry in the West in the second half of the nineteenth century.

Hoagland selected Forts Laramie, Bridger, and D. A. Russell (today F. E. Warren Air Force Base) because of their unique contributions to America's western story, and because so much original architectural fabric survives at each place. As a military post, Fort Laramie (1849–1890) figured in virtually every dimension of the western movement, including the mid-century pioneers and argonauts trekking west, the Pony Express and telegraph, Indian treaties and campaigns, and the Black Hills gold rush. Fort Bridger (1858–1890) shared some of Laramie's story, but not its geography, substituting the vagaries of mountain and basin for Laramie's timberless plains. Fort D. A. Russell (1867–1930; Fort Francis E. Warren, 1930–1949; F. E. Warren Air Force Base, 1949–present) was founded to protect the advance of the nation's first transcontinental railroad, but owes its remarkable longevity to Francis E. Warren, cattleman, governor, and then U.S. senator. Warren was a towering politician with an insatiable and aggressive knack for tapping the federal trough. Senator Warren represented Wyoming in the U.S. Senate for thirty-seven years (1890–92, 1894–1929), variously chairing the Military Affairs and Appropriations Committees. During the era of fort abandonment and consolidation in the late nineteenth century, Warren saw to D. A. Russell's continued existence and structural expansion, expanding the garrison to brigade size early in the new century.

Critically important to Hoagland's study, elements of Laramie's, Bridger's, and D. A. Russell's architectural legacies have been preserved by their modern-day owners, and she dissects these elements like a surgeon. The first army building constructed in Wyoming in 1849, for instance, survives today at Fort Laramie National Historic Site, jealously protected by the National Park Service. Remarkable buildings survive at Fort Bridger, too, while at Fort D. A. Russell numerous buildings from the army's golden age of brick construction survive.

Hoagland's story is wonderfully complex and deftly woven. Rhyme and reason can, indeed, be discerned in the actions of local post quartermasters as they laid out these posts and expanded and transformed them. Hoagland makes a convincing case that fort designers knowingly embraced and replicated stereotypical New England villages, whose central greens and orderly layouts were deeply respected and emulated in the East. Of course, posts needed central parade grounds for mounting guards, mustering troops, and serving other ceremonial and functional purposes, but this utilitarian need could as easily have been positioned on the margins instead of at the center of an army structural layout. Indeed, central parade grounds, towering flagstaffs, and orderly buildings made powerful statements to onlookers about the civilizing forces at work wherever the army planted itself.

Laramie's and Bridger's architecture might be labeled eclectic at best. No two buildings were much alike. Outmoded and modern architectural styles stood side by side, as did buildings constructed of log, lumber, adobe, stone, and, at Fort Laramie, concrete. After numerous unsuccessful attempts, the army did at last embrace standardized plans for almost every type of building conceivably needed at its posts, thus giving rise to a golden age of construction at posts that well outlasted the nineteenth century and is plainly evident at installations across the land today. By then the army's strategic thinkers had eliminated dozens of small, aged posts scattered hither and yon across the western landscape in favor of large centralized forts, stoutly built, and located along railroad lines so as to enable the rapid movement of troops to places of need.

Consolidation gave rise to a handful of brick forts scattered across the nation. Although Fort Laramie, for example, featured several buildings constructed of stone and many more of concrete, its importance gradually diminished and the post was abandoned. It had no brick construction. Bridger featured some buildings constructed of stone, but it, too, was abandoned as the frontier passed on. The same was true for many other larger northern plains posts, such as Forts Custer, Keogh, Buford, Sully, and Randall. But Fort D. A. Russell had a neighboring capital city, Cheyenne, a position on the Union Pacific Railroad, and the dogmatic and tireless Senator Warren. In due course it was transformed from its rustic frontier progenitor to a lavish, nearly all-brick complex that kept growing in stature and size well into the twentieth century. Though functioning today as an Air Force base whose airmen serve missiles, not planes, it has no runways but does showcase, among other massive red-brick buildings, twenty-two surviving, nearly identical, all-brick cavalry and artillery stables.

Hoagland's close examination of the architectural nuances of Forts Laramie, Bridger, and D. A. Russell has applicability across the military West. Individual buildings and appointments had parallels on the northern plains, whether at Fort Buford, Assinniboine, McKinney, Niobrara, or Meade, and certainly elsewhere in America. Posts met local needs, were adapted to local conditions and materials, and variously were expanded, abandoned, and succeeded. Until now, however, scholars of the U.S. Army and its physical complexes have not enjoyed such a thoughtful and detailed analysis as this. Alison Hoagland's *Army Architecture in the West* is a significant contribution to the legacy of the Old Army and one of its most fundamental institutions, the forts where troops served, in Wyoming, in the West, and across the land.

PAUL L. HEDREN

ARMY ARCHITECTURE IN THE WEST

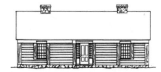

INTRODUCTION

When beginning this study of military architecture in the American West, I assumed that I would be writing the story of imposing structures that commanded attention and evoked brute power.[1] Forts are forts, after all, and as such they should be fortified places representing military might. A painting such as Charles Schreyvogel's *Defending the Stockade* (ca. 1905) captures the popular view of army architecture in the West, which I shared: high, palisaded walls enclosing a defensible stronghold (fig. I-1). I took for granted that these forts would also have a certain uniformity throughout the region, since people in Washington were responsible for their layout and design. I was wrong.

Visits to forts in Wyoming and elsewhere revealed them, first of all, to be very different from each other, neither monolithic nor uniform. And they were rather unimposing in architectural appearance. Even taking into consideration that only a small number of nineteenth-century buildings survive at each fort, the sites seemed small—several dozen buildings huddled together on a wide, open landscape. High walls, blockhouses, and fortifications were usually absent. As one observer noted, "A plains fort is no fort at all . . . no bastions, walls, stockade, nor defence of any kind."[2] These forts did not look fearsome, powerful, or intimidating in the least. Clearly, the forts were not the forts of my imagination.

Another reading of western fort architecture was suggested by a photograph of a woman on a bicycle at Fort Meade, South Dakota, taken in 1897

Fig. I-1. "Defending the Stockade," painted by Charles Schreyvogel, ca. 1905. Images such as this popularized the notion of western forts as having stockaded walls. From Truettner, ed., *The West as America*, 304.

(fig. I-2). The many incongruities in this image—of a woman in a seemingly male world, of a bicycle in a landscape animated by horses, and of eastern civility transplanted to the western frontier—rendered the stockaded fort stereotype useless. Furthermore, as my research progressed, I realized that this photograph was even more telling than I had known, for contained in it was the very paradox that *is* the western fort. American military installations in the West were rugged, remote, and dangerous places, certainly, but they also had a remarkably genteel side, with visible connections to eastern cultural sensibilities. The presence of a bicycle on a boardwalk is no less surprising than many other, more obvious efforts to connect the forts to the East, including the use of architecture. The cupola-topped bandstand in the background of the photograph is no whimsical vagary. Military planners used such elements as clapboard siding, shingle roofs, bay windows, neoclassical symmetry, and specialized floorplans—as well as gardens, trees, bushes, and picket fences—to fashion a landscape suitable for officers and also, though perhaps less self-consciously, to represent visually the interests of the dominant culture in this new American territory.

Designs of both the posts and the buildings that composed them owed much to military tradition. Posts had a common design, featuring an open parade ground framed by barracks and quarters in a rectilinear arrangement. The buildings were similarly rectilinear, usually in a neoclassical style that valued symmetry and order. In the post–Civil War period, picturesque ornament often cloaked symmetrical plans. The quartermaster general's office in Washington, D.C., suggested designs, but officers at the division, department, and post level usually produced them. While the same design might appear repeatedly at a post, designs were not shared among posts, so that each post retained its individuality. Further, the original design of a post maintained its coherence for only a few years, until new buildings were added that introduced a new style or material or that violated a rational layout.

Architectural design was only part of the story; the process of construction explains the disparity between what was planned and what was actually built. Located at the end of long supply lines, thousands of miles away from the nearest architect, subject to the approval of a chain of officers of varying expertise and interest, the forts suffered a variety of influences and constraints on their construction. Ideals from the East mingled with customs in the West as army builders confronted a land that largely lacked their most familiar construction material. Builders at the posts invented solutions, borrowed ideas, adapted published designs, and ultimately ceded authority to Washington, but only after a long period of improvisation. Additionally, a dwelling's furnishings and fixtures—providing heat, light, and water—affected how that building was used and understood. Examining how these new technologies were introduced into barracks and quarters reveals a workforce that alternately rejected and adopted, improvised and turned to professionals. This process of developing an architecture, which ranges necessarily from public image to construction techniques, is the subject of this book.

Precedents for these western forts were rare. When the nation was founded, forts served to defend seaports along the East Coast. Military engineers designed high masonry walls and bastions to permit firing at hostile ships approaching key harbors and to withstand shelling. The first inland forts of the young nation followed the same form, sometimes in stone, but also in wood—a high, stockaded wall with blockhouses at the corners. As the defensive line moved west in the 1820s and 1830s, however, military strategists slowly realized that high walls were a large and unnecessary expense; Indians rarely attacked fortified positions.[3] Thus, the western fort was an evolving form, one without proven effectiveness or clear precedent.

Fig. I-2. Parade ground, Fort Meade, South Dakota, 1897. Although a woman on a bicycle is a seemingly incongruous user of a parade ground, the open space and boardwalk make it an ideal setting. Courtesy Library of Congress.

The novelty of the situation enabled post commanders to experiment, within certain limitations.

The army had several objectives in the West, both overt and unstated. Its primary goal was to exert military power, a mission that Gen. William T. Sherman described as "the great battle of civilization with barbarism." For the general, and for most Americans of the time, "civilization," the highest evolution of a society, was defined by its opposition to savagery, which was represented by the Indians. Military conquest was the first step in civilizing the savage. Between 1848 and 1890, the U.S. Army engaged in more than twelve hundred combat actions against Indians. In the 1850s, four-fifths of the entire U.S. army of about ten thousand soldiers was stationed west of the Mississippi; in the 1890s, about two-thirds of the army of twenty-five thousand were stationed in the West.[4]

The army also performed an important function for eastern travelers and settlers. The army initially came west to aid emigrants traveling the overland trails to the West Coast; containing the natives was a secondary but related mission. Even after completion of the transcontinental railroad, travelers continued to pass by and through the forts, like the thousands who passed by Fort Laramie in 1875–1877 on their way from the railroad in Cheyenne to a gold rush in the Black Hills of South Dakota.[5] Travelers stopped at the forts for supplies, relied on them for protection, or at the least saw them as landmarks on their journey.

The western army also played an important role in a kind of economic conquest, putting dollars from Washington directly in the hands of western settlers. Neighboring settlers saw the fort as a marketplace, visiting it frequently to buy or sell goods and services. Much as they do today, forts granted many contracts to civilians for the provision of necessities such as hay, wood, and other consumables. Settlers in need of cash also gravitated to the fort for short-term employment. While the army often imported skilled workers from more populated areas, forts also employed local people for construction and transportation duties. In 1867 Fort Laramie carried 300 civilians on its payroll to complement 472 officers and soldiers. Settlers did not view forts as remote, Indian-focused establishments outside their realm, but rather as accessible resources, there to serve them, buy from them, or hire them. The economic relationship continued into the twentieth century, when Fort D. A. Russell acquired the sobriquet "a monument to the pork barrel" through its $5 million rebuilding program that benefited local contractors.[6]

As other historians have noted, conquest is also a cultural process, and accordingly, the architecture of forts expressed the army's implicit cultural aspirations. Expressions of culture incorporate common ideas about an individual's relationship to society, and the built environment is a particularly rich medium for understanding these connections. Buildings constitute a tangible expression of how an individual relates, or is intended to relate, to his or her peers, superiors, family, and the world at large. Because architecture shapes these relationships, an analysis of the architecture helps give an understanding of people's lives and attitudes. Particularly for people who do not leave written documentation of their ideas and beliefs, architectural evidence can be revealing. And even for those who do document themselves self-consciously, such as army officers, their buildings can betray them.[7]

In the case of the western army, the architecture allows us to see what ideology the army promoted, consciously or not. While army officers may have been writing the rhetoric of centralized command and uniform appearance, their posts reveal flexibility and even idiosyncrasy. In addition, the architecture reveals the struggles of a geographically sprawling army to achieve a standard image and reflects efforts to innovate within a poorly funded bureaucracy. Army architecture was, at times, shabby and genteel, rudimentary and stylish, domestic and military. Like the army that produced it, the architecture reflected conflicting goals and competing interests.

In the last half of the nineteenth century fort builders attempted to incorporate an attitude of superior refinement in the architecture. Western army officers were keenly aware that they represented civilization to settlers and emigrants. Officers not only outranked soldiers, they were better educated than their men, better trained, better clothed, and better behaved, even "gentlemanly." Richard Bushman has explored the concept of "gentility" among the American upper class in the eighteenth century as it evolved into aspirations of "respectability" among the middle class in the early nineteenth century. He found these ideas embedded in the bourgeoisie's new-found affinity for neoclassical architecture. Similarly, army officers of the mid- to late nineteenth century were raised with middle-class ideals of refinement, ideals that were reinforced in their training The officers expressed these aspirations through architecture in the three hundred western forts built by the army in the nineteenth century. That officers and soldiers slept in beds and bunks, in buildings with floors and windows, constituted evidence in their own minds of the superiority of their civilization over that of peoples who slept on dirt floors in windowless lodges. Further improvements, such as bay windows and clapboard walls, were not so much directed at expressing superiority to Indians as at displaying an eastern cultural preference for refinement. Margaret Carrington, an officer's wife, wrote that the wood shingles made by the troops at Fort Phil Kearny served to "convince the skeptical that shingles, or anything else, can be made or *done*, when it *had* to be, and that civilization was still westward bound," explicitly relating architectural improvements to her civilizing mission.[8] Gold seekers along the Bozeman Trail might be impressed by the refined appearance of the remote post, but its finished quality was also a concern to Carrington for her own satisfaction. Her definition of herself as a successful officer's wife rested on her creation of an appropriately genteel domestic environment, which reflected a memory of the culture she had left behind.

The decision to narrate the story of the building of the western military landscape by examining the architecture of three Wyoming forts requires some explanation. Although forts in the Southwest followed a different pattern, borrowing from long-established building traditions present there, Wyoming forts were typical of many in the northern Great Plains, and all forts were products of the same system. Wyoming was central to the army's occupation of the West, as it was the territory traversed by the main route to Oregon, California, and Utah, and also received the first transcontinental railroad line. Protection of these routes was the original reason for the army's intrusion into the West. Establishing a presence over as much of the region as possible accounted for the army's construction of hundreds of posts; after the Civil War, the army launched offensive actions against the native population from many of the posts, including those in Wyoming.

Of the seventeen forts built in Wyoming before 1898, the three posts that were garrisoned the longest and that have the greatest number of buildings surviving from the nineteenth century serve as the basis for this study: Forts Laramie, Bridger, and D. A. Russell.[9] Each had a different reason for being. In 1849 the army acquired Fort Laramie, located near the Nebraska border on the Oregon Trail, from a fur-trading company in order to protect increased emigration along that trail. In 1858 the army seized Fort Bridger, originally a trading post and then a Mormon post, and maintained it both to keep a military presence near the Mormons and to protect the southern overland trail to California. In 1867 the army built Fort D. A. Russell adjacent to Cheyenne, locating the fort along the Union Pacific Railroad line in order to facilitate its construction. The decision to study these three forts proved fortuitous. With different origins, building materials, layouts, and histories, the forts varied considerably while still being within a single geographic area and administrative unit—Wyoming and the Department of the Platte, respectively—which gave a certain cohesion to the project. Familiarity with decades of annual reports of the secretary of war and publications such as the *Army and Navy Journal,* along with a cursory examination of other western forts, led me to conclude that these three were representative in their individuality. They did not look like each other, and others did not look alike, either.

In the West, fort building evolved over the last half of the nineteenth century and the early years of the twentieth. In the first phase, the fort was an outpost, remote from cities and at the end of long supply lines. Initial attempts to fortify these posts with high walls or other defensive structures were often soon abandoned, and the forts often took the form of an open, irregular collection of buildings oriented around a parade ground. The construction of forts faced numerous constraints: untrained personnel, ranging from troops as construction labor to lieutenants as designers and construction supervisors; inadequate space allotments, with the number of rooms for officers determined by rank, regardless of family size; and shortages of construction materials so that, in the absence of readily available wood, army builders frequently used adobe. The tension between military ideals and local tradition produced forts that achieved recognition as a familiar form, yet tended to be more disorderly than their planners intended and shabbier than their occupants desired.

In their second phase, western forts were commonly likened to villages by contemporary observers. Physically, the fort was organized around an open space, much like a village green; the buildings were a mixture of styles and materials, much as if they had been built by private citizens over several decades; lesser buildings appeared in a disorderly arrangement, away from the center; and trees, porches, sidewalks, and gardens contributed to the domestic atmosphere. The village-like appearance of forts reflected a desire by post commanders to retain control over design and construction, even as the army was increasingly able to impose a uniform image through improved transportation systems and a centralized design process. Thus, model designs issued by the quartermaster general were adapted, not adopted, at the posts. Efforts at standardization succeeded more readily with hospitals and nonresidential buildings than with officers' quarters and barracks. But the increasing complexity of buildings eventually outstripped the abilities of untrained personnel at the posts: plumbing and heating were beyond the post quartermaster's ken, and once Congress required that construction be contracted to civilians, post quartermasters' inadequate drawings and specifications caused them to rely increasingly on those produced by the quartermaster's department in Washington.

In the third phase, the informality of the village gave way to the modern fort, in which the army presented itself as an institution. The changes begun before 1890 took root, and architecture became standard, construction became professional, and a uniform appearance spread from post to post. The idiosyncratic, village-like forts were rebuilt or replaced as the army moved into the modern era. Monolithic forts are the image of the army with which modern Americans are familiar, but they came rather late in the western army's fort-building history.

The subsequent history of all three posts explains why so many nineteenth-century buildings survive into the present. Fort D. A. Russell remains a military post, now known as F. E. Warren Air Force Base. Fortunately, new construction has occurred outside the historic core, leaving the heart of the fort much as it was on the eve of World War I. Forts Laramie and Bridger are both open to the public. Site administrators are engaged in the task of interpreting the surviving buildings to tourists in a way that helps them understand the past. This book offers a new way of using the architecture to gain insight into the army's attitude toward its role in the West, examining not only why buildings looked the way they did, but also how they came to be built. A dynamic combination of military needs, local tradition, and eastern U.S. culture influenced the choices made in the construction and maintenance of army buildings.

PART I | OUTPOST
Forts Laramie and Bridger, 1849–1869

1

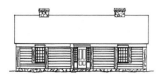

AN OASIS IN THE DESERT

Purpose and Plan

On November 14, 1849, a group of Church of Latter Day Saints missionaries, heading east from Salt Lake City, reached Fort Laramie. One of them reported, "On our arrival at Fort Laramie we obtained supplies for our selves and horses . . . There is an air of quietness and contentment, of neatness and taste, which . . . made us feel as if we had found an oasis in the desert."[1] Although these Mormons were heading in the opposite direction of the stream of travelers who passed by Fort Laramie, their observation contained two characteristic elements: they received supplies at the fort, and they were struck by its air of civilization (fig. 1-1).

Attracted by the discovery of gold in California, in 1849 gold seekers augmented the already considerable traffic of Oregon-bound settlers and Utah-destined Mormons, to the extent that estimates of travelers on the westward journey that year ranged from twenty thousand to forty thousand. Most of them passed by Fort Laramie, which the U.S. Army had acquired as a military post on June 26, 1849, as the fort occupied an important site on the trail that later diverged to Oregon, Salt Lake City, or California (map 1-1). Most of the travelers used the facilities at Fort Laramie, just as they had when it was a trading post. A soldier observed, "Nearly all the parties remain here a few days to reset wagon tires, exchange and purchase cattle, mail letters for the states, and replenish their supply of provisions from the commissary, who is permitted to 'sell to those actually in want.'" As the emigrant throng continued—the register at Fort Laramie in 1850 recorded nearly ten thousand wagons—visitors often remarked on

Fig. 1-1. Fort Laramie, 1858. The fort may have earned the description "an oasis in the desert," but this early photograph shows a muddy, scattered settlement. The adobe trading post of Fort John still stands, on the left, and Indian tepees are visible in the foreground. Courtesy the Lee-Palfrey Family Papers, Manuscript Division, Library of Congress.

the quality of the architecture, the number of buildings, and the facilities available when they reached Fort Laramie.[2]

These thousands of emigrants were the initial reason for the establishment of forts in the West. Protection of the emigrant trails implied conflict with the Indians, and whether the army's role was safeguarding travel routes or actively pursuing "hostiles," its mission was entwined with the Indians. The army had other, related roles in the West: exploring and mapping, building roads, facilitating railroad construction, and controlling Mormons, to name a few. But it was confrontation with the Indians that framed the fort-building strategy for the next forty years. The nature of the army's conflict with the Indians changed as the army's goal shifted from protecting emigration to facilitating the process of settlement. U.S. government relations with the Indians changed as well, as government

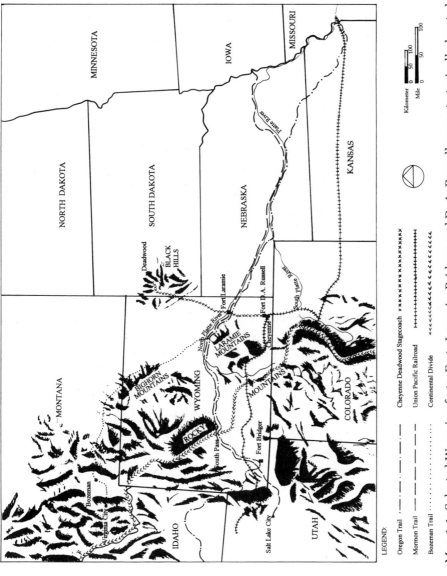

Map 1-1. Selected Wyoming forts. Forts Laramie, Bridger, and D. A. Russell were strategically located on transportation routes. Drawing by Julie Osborne, 1996, Western Regional Architecture Program, Graduate School of Architecture, University of Utah.

policy shifted from conceiving the land as Indian territory, with posts and trails within it, to confining the Indians to defined spaces and considering all other land available for settlement and traffic.

The first phase of army policy in the West—facilitating emigration—produced a distinct type of fort with its own kind of buildings. Conceiving its role as essentially defensive, the army planned fortifications in the construction of its first forts. The fortifications, usually in the form of walls, indicated that the army saw its position as exposed and vulnerable, but high, impervious walls also served to convey an image of military might. Within the fortifications, rectilinear buildings arranged in a neat and orderly fashion would impart an appropriate message of military discipline. But these ideals faced a difficult translation into actual fact. Inadequate funding and supply systems combined to produce forts that were more likely unfortified, haphazardly arranged, and poorly constructed. While the forts may have been less military in appearance than desired, they nonetheless contrasted sharply with Indian settlements and with homesteaders' ranches. Their size and basic arrangement managed to convey a sense of civilization—of East Coast refinement set out on the Great Plains.

Purpose

The army's mission in the West was key to the kind of forts that it would build, and that mission was tied to U.S. policy toward the Indians. In the 1830s federal policy makers entertained the idea that most of the land west of the Missouri River would be of little interest to settlers, and therefore would be appropriate for Indian territory; removals of Indians in the southeastern United States to this region were predicated on this belief. A congressional act of July 2, 1836, called for the "better protection of the frontier," and the resulting Poinsett Plan proposed a north-south line of forts, defining the "permanent Indian frontier." By 1846, however, the concept of a "permanent Indian frontier" had been shattered. In that year, the United States acquired the Oregon Territory, and in 1848, with the conclusion of the war with Mexico, it gained California, Texas, and land between, ensuring that a steady stream of emigrants would be crossing the Great Plains in order to reach the West Coast.[3]

Indian tribes that traditionally occupied these lands—including the Sioux, Cheyennes, Arapahos, Crows, Assiniboins, Gros Ventres, Mandans, and Arikaras—objected. Emigrants depleted game in the lands adjacent to their travel routes, game that the Plains Indians also depended upon. The Indians, in turn, raided the emigrants' trains in order to obtain livestock. Accordingly, the United States negotiated a treaty with these northern and central Plains Indians in 1851. Known as the Treaty of Fort Laramie, the agreement assigned specific territory to each tribe, and held that tribe responsible for attacks on emigrants within its boundaries. The treaty asserted the right of the federal government to establish military posts and roads. In return, it promised annuities—annual payments of goods—as compensation to the tribes for ten years. Enforcement of this treaty, especially ensuring the protection of the emigrant trails, fell to the U.S. Army. Emigrants demanded protection, and congressmen satisfied their constituents by authorizing fort construction.[4]

Tensions heightened in 1854 with the incident that is known as the Grattan Massacre. A Sioux Indian slaughtered a cow that had wandered from an emigrant train, and the emigrant owner lodged a complaint with the army at Fort Laramie. Young Bvt. 2nd Lt. John Grattan took a force of thirty men to the nearby Brule Sioux camp and demanded that they turn over the culprit. Although the Sioux offered restitution, Grattan ended discussions and ordered his men to fire. The Sioux fought back, killing all of the soldiers except one, who made it back to Fort Laramie before he died. The next summer, Col. William S. Harney marched west from Fort Kearny, Nebraska, to retaliate. At Blue Water Creek, Nebraska, Harney attacked a Brule Sioux village, killing and imprisoning more than half of the occupants.[5] These hostilities began an intermittent war on the Great Plains that lasted several decades.

Through the 1850s, forts such as Laramie served as staging grounds for expeditions against the Indians. Excursions included campaigns (in which soldiers were accompanied by supply wagons), patrol actions (military pursuits instigated following news of some sort of depredation), and escort duties (in which a few men were assigned to accompany persons or a cargo).[6] Most of the soldiers' time was occupied with fatigue duty—the routine chores, such as obtaining firewood and water, that facilitated daily existence. The provision of shelter, fuel, and food required soldiers' ongoing attention.

Also through the 1850s, the emigrant treks continued. In 1852, forty thousand people are estimated to have passed by Fort Laramie, although only about half that number made the journey the next year. The Mormon emigration to Utah was especially heavy, with some three thousand Saints pushing handcarts; another five thousand traveled west in wagons in the late 1850s. Freighting along the Oregon-California trail, much of it to supply military posts, was also considerable in this period. Overland freighters Russell, Majors & Waddell, who obtained contracts from the military for transporting goods, employed four thousand teamsters in 1858. In 1860, twelve thousand men freighted for various outfitters on the Plains.[7]

With the outbreak of the Civil War, much of the Regular Army shifted east, and volunteer forces backfilled at western posts. As traffic continued on the emigrant road, so did Indian hostilities. The Shoshonis and Utes perpetrated a series of attacks on the mail route west of Fort Laramie; the agent holding the mail contract received permission to move the mail route 100 miles south of the emigrant trail through Wyoming, crossing the divide at Bridger's Pass instead of South Pass. In Colorado, Col. John M. Chivington led a force of 700 men against a village of Cheyennes at Sand Creek, killing about 200, two-thirds of them women and children. The outrageousness of this act did not escape the American public; even the military could not defend Chivington's actions. The attack at Sand Creek further escalated the hostilities, with Indian retaliations and army reprisals occurring across the Plains in 1865.[8]

After the Civil War, U. S. Grant appointed William T. Sherman to head the Military Division of the Missouri, a vast expanse that swept from the Mississippi River to the Rocky Mountains, and from Canada to the Gulf. Sherman's initial strategy was to protect two or three routes through this territory, without attempting to guard every individual settlement. His instructions to Brig. Gen. Philip St. George Cooke, commanding the Department of the Platte, were: "Your first and imperative duty is to protect the Telegraph and Stage lines which lie along the Platte, to Utah, to Colorado, and Montana."[9]

Sherman vs. Meigs

The rational location of posts and the construction of sound buildings was not helped by the ongoing conflict between the two men at the top, Commanding Gen. William Tecumseh Sherman, who served from 1869 to 1883, and Quartermaster Gen. Montgomery C. Meigs, who served from 1861 to 1882. In the line-and-staff organization of the army, Sherman reported to the president and Meigs to the secretary of war; Meigs did not have to answer to Sherman, nor did Sherman have to take Meigs's advice.

At the close of the Civil War, Commanding Gen. Ulysses S. Grant appointed Sherman, the hero of the Union Army's brutal march through Georgia, to be head of the Military Division of the Missouri, which included all of the Great Plains in its territory. When he was elected president, Grant appointed Sherman Commanding General of the Army. Sherman immediately observed that the staff had become independent of the line; by 1873 he groused that "no part of the Army is under my direct control." In 1874, disgusted with the situation and in conflict with the secretary of war, William W. Belknap, Sherman decamped to St. Louis, taking the headquarters of the army with him, for nineteen months.

Meigs also found fault with his own position. He had graduated from West Point in 1836 and initially entered the Engineer Department. In the 1850s he oversaw construction of the aqueduct in Washington, D.C., as well as the wings and dome of the Capitol, before taking over the Quartermaster Department at the outbreak of the Civil War. His dissatisfaction with his job after the war stemmed from the fact that he had few men under his control. Post quartermasters were line officers, who reported to post commanders, not staff officers in the quartermaster's department. Orders from the quartermaster general's office to post quartermasters were routed through the assistant adjutant general at department headquarters and the posts' commanding officers.

In 1866 Meigs objected to the selection of the site for Fort Sedgwick, in northeast Colorado, on the grounds that it was an extremely expensive post to supply because of a lack of timber within a reasonable distance. Sherman was furious, insisting that site selection was his prerogative. He fumed to the commanding general, U. S. Grant, "The Quartermaster General construes himself the judge of what improvements shall be made. This is all wrong. We who command troops must station them, and we must be the judge of the kind of structures needed."

The controversy heightened in 1867 when poor site selection forced the army to abandon several partially built forts in Texas. The quartermaster general detailed funds squandered through poor planning, including $7,000 spent on Fort Belknap before the commanding officer recommended that it not be occupied, $41,000 spent on Buffalo Springs before it was determined to be "inadequately supplied with water," and $37,000 spent on Fort Chadbourne before it was abandoned due to "the insufficiency and unhealthy quality of water." Despite the inefficiencies, the struggle over fort placement continued.

The disagreement also extended to building construction. In 1866 Sherman obtained a $1 million appropriation—diverted from Meigs's account by order of the secretary of war—for construction of forts in his Division of the Missouri. In administering this "sheltering fund," Sherman retained final approval of all construction, and his chief quartermaster at the division level disbursed the funds, sending $300,000 to the chief quartermaster of the Department of the Platte for distribution to posts there. Sherman assured Brig. Gen. P. St. George Cooke, commanding the Department of the Platte, "I will endeavor so as to shape things that the building of posts . . . will be left to you and your staff as much as possible," but then cautioned him that "economy of the meanest kind, save when life is involved, must be the very first consideration. . . . If Department Commanders exaggerate their wants so as to appropriate above the limit allowed by law, we cannot complain that General Meigs should insist on estimates and his approval beforehand." This "sheltering fund" was an affront to Meigs; he complained that he didn't know how the money was being spent.

> In another incident the next year Meigs disapproved new construction at Fort Larned because of the inadequate drawings supplied by the post. Sherman again complained to Grant: "My quartermaster [at the division level] should have supervision over the affairs of the quartermaster department in all the country subject to me. . . . Then you could set aside to me what you could spare to enable me to house my troops. Knowing the necessity of service I could apportion the amount to my several departments, discriminating against those who have timber and materials that soldiers can use, and favoring those that must have money." Not surprisingly, Meigs saw the situation differently, arguing that the staff was responsible for expenditures, and that to turn funds over to division commanders to "spend them at your will" would be irresponsible. The line-and-staff conflict, so vividly demonstrated by the bickering between Sherman and Meigs, and echoed by their subordinates, inevitably affected the design of army posts as well as their placement.
>
> Sources: *Annual Report of the Secretary of War* (1869), 29; (1873), 26; (1868), 855–871; (1866), 4; (1867), 531. Weigley, *Quartermaster General of the Union Army*, 350–351. *Army and Navy Journal* 16 (25 January 1879): 431. Sherman to Grant, 22 June 1866, RG 393, Division of the Mississippi, Letters Sent, 4:405, NAB, cited in Sheire, "Fort Larned National Historic Site, Historic Structure Report, Part II: Historical Data Section," 14. Risch, *Quartermaster Support of the Army, 1775–1939*, 485. Sherman to Cooke, 22 October 1866, RG 393, Part I, Entry 3731, Box 1, NAB.

A gold discovery on Grasshopper Creek, Montana, spawned traffic along the Bozeman Trail, which diverged from the Oregon-California Trail west of Fort Laramie and headed north through the Powder River country of Wyoming. To guard travelers, in 1866 Sherman authorized construction of three forts along the Bozeman Trail. Col. Henry B. Carrington placed his headquarters at the new Fort Phil Kearny, about halfway between Fort Laramie and Bozeman. Sioux Indians, loosely organized under Red Cloud, harassed soldiers building the forts. Although they never attacked the well-fortified posts directly, they did attack a wood train from Fort Phil Kearny in December 1866. When Capt. William J. Fetterman led his soldiers in retaliation, apparently against orders, the Indians killed all eighty of them.

While the army pledged reprisals, other forces in Washington saw this as an opportune time to press for peace. A peace commission, of which Sherman was a somewhat reluctant member, sought agreements with the Plains Indians. As part of the Fort Laramie Treaty of 1868, the army abandoned its Bozeman Trail forts; Sherman claimed they were not much of a loss, as the newly built railroad opened up a more western route to the gold fields. Signed by Red Cloud in 1869, the treaty created the Great Sioux Reservation, which comprised approximately the western half of present-day South Dakota, and caused a shift in the army's role in the West.[10]

The army's early mission of protecting trails resulted in a policy of numerous small, fixed posts, arrayed along these trails. Secretary of War John B. Floyd articulated this policy in 1857, although he still perceived the frontier as a line: "A line of posts running parallel with our frontier, but near to the Indians' usual habitations, placed at convenient distances and suitable positions, and occupied by infantry, would exercise a salutary restraint upon the tribes, who would feel that any foray by their warriors upon the white settlements would meet with prompt retaliation upon their own homes." The alternative favored by the previous secretary of war, Jefferson Davis, Commanding Gen. Winfield Scott, and Quartermaster Gen. Thomas Jesup involved concentrating the forces at fewer, larger posts, and sending roving columns of troops through the Indian country. Davis explained that this would "restrain aggression by the exhibition of a power adequate to punish." The expense and difficulty of supplying the small posts had to be weighed against the security they offered to emigrants and the presence they showed to Indians. The public, comforted by the proximity of the military, favored the small posts. When Davis attempted to concentrate his forces, he found "the want of troops in all sections of the country so great, that the concentration would have exposed portions of the frontier to Indian hostilities without any protection whatever."[11] The area was too vast to permit concentration.

For the next several decades, the various commanding generals and secretaries of war explained to Congress repeatedly that this strategy of saturation had a cost: supplying scattered posts was expensive. Sherman, in his first report as commanding general of the army, noted that he could concentrate his troops near supplies, or "if we had, as in former years, a single line of frontier a little in advance of settlements," his job would be easier.[12] As conditions stood, these posts required costly support, in part because they occupied land where fuel and building materials were not readily available.

The army's mission to protect emigrant trails had another effect: the numerous small forts would be temporary in nature, not permanent. The vision of the frontier, or the edge of civilization, as a moving line meant that forts would have to move to keep just ahead of the frontier. But even after the view of the frontier as a north-south line had faded, the army continued to see forts as a temporary aid to settlement. After five, ten, or fifteen years, when the country around the fort had become sufficiently settled to discourage Indian aggressions, the army would abandon it and move on. Nearly all of the posts constructed in the West were temporary, even when they should have been permanent. Congress ruled in 1872 that it had the right to approve any permanent post whose cost exceeded $20,000, resulting in a virtual ban on permanent posts. Lt. Gen. Philip Sheridan pointed out, "This necessitates a waste of money for temporary shelters, which could otherwise be used in the construction of permanent posts."[13]

The designation of posts as temporary meant that they would be designed by personnel at the post or at the department level, that troop labor would construct the buildings, and that impermanent materials of wood or earth—not stone or brick—would be used. These three factors combined to produce substandard dwellings, and Sherman felt the inadequacy of the accommodations keenly. After touring western posts in 1866, he railed against the quality of the buildings, which he routinely termed "huts," in his appeal for greater funding. He wrote from Washington in 1869: "I have personal knowledge that the huts in which our troops are forced to live are in some places inferior to what horses usually have in this city."[14] While the temporary nature of western posts may have been strategically necessary in some cases, the designation of all posts as temporary resulted in a generally shabby appearance of constructions that were supposed to be expressions of U.S. military might.

Temporary construction also resulted from insufficient appropriations. Despite its generally favorable press, the frontier army did not have the overwhelming support of Congress. The wariness of a standing army expressed in the Bill of Rights persisted through the nineteenth century. Legislators steadily reduced army appropriations and personnel in the years when its frontier role was expanding. In the 1850s Secretary of War Jefferson Davis was able to win some modest increases by pointing out that in the previous fifty years the nation had doubled in size, the white population had grown by 18 million, yet the strength of the army had remained at about 10,000. In general, though, antebellum congresses were reluctant to spend money on increasing the size of the army, or even on adequately supplying the existing army. After the Civil War, demobilization was the initial concern; almost a million volunteers had to be mustered out. The strength of the Regular Army, which Congress regulated by controlling the number and size of units, was set at about 54,000 by law of 1866; this was reduced to 30,000 in 1870. In 1874 it was further reduced to 25,000 enlisted men, where it stayed until the Spanish-American War. Because of the lag in filling vacancies, the actual number of enlisted men rarely exceeded 19,000. Historian Robert Utley cites 1881 as a typical year, in which the actual strength of a cavalry troop averaged 58, including 46 privates, and of an infantry company, 41, of which 29 were privates. With the 430 companies of the army spread to cover more than 200 posts, most Western posts had only 2 or 3 companies, or less than 100 men. Furthermore, they were mixed garrisons, with troops of cavalry for active duty and companies of infantry to guard the fort or temporary field camps.[15]

The best evidence of the low esteem in which the public held the army was the quality of architecture of the forts. Although the army had adopted a policy of fixed forts to accomplish its varied missions in the West, it did not have the support of Congress to furnish these forts in a manner that inspired pride in the troops and instilled fear in its enemies. Instead, the forts were minimal constructions of locally available materials, erected with the dubious skill of the troops. The forts were also placed idiosyncratically.

Site

Once military commanders had made the decision to build a fort to support a certain mission, the next decision was where to place it. Commanders at the department, division, or headquarters level decided that a fort should be located along a certain trail, at a judicious distance from the previous fort, for instance, but the actual location was in the hands of the commanding officer. In 1858, Quartermaster Gen. Thomas S. Jesup outlined the "three important elements which should govern in the establishment of a military post" in the West: "They are, first, an abundant supply of pure water; second, wood for fuel and for building; and, third, grass for the support of public animals in summer, and the supply of hay for them in the winter. If either be wanting, not a cent should be expended on any interior site, no matter what its supposed military advantages in other respects may be."[16] The availability of water, wood, and grass—especially wood—could not always be assured, resulting in soaring costs for supplying the western posts.

Critics perceived an almost whimsical approach to site selection. The *Cheyenne Daily Leader* mocked Col. Henry B. Carrington's placement of Fort Phil Kearny, alleging that after several days' march, his wife, "a lady of some will," refused to move on one morning and declared that this is where the new fort would be. The paper criticized the selected site as "a most exposed and defenseless place, commanded by bluffs on every side." In fact, Carrington and his officers examined several possible sites before staking out the lines of the post, and a senior officer praised Carrington's "excellent judgment in selection of position." The newspaper's attack perhaps reflected Carrington's unpopularity more than his ineptitude.[17]

The placement of army forts was neither as serendipitous nor as calculated a process as it was variously portrayed. The establishment of two Wyoming forts revealed a clear sense of mission and a willing acceptance of preexisting facilities. Fort Laramie was a functioning trading post when

> ### Location of Posts
>
> In 1882, Lt. Col. Thomas M. Anderson expressed his views on the haphazard location of many army posts:
>
>> If our posts are occasionally found to be injudiciously located, it is often from the fact that the position taken for them was the result of a mere chance combination of circumstances, no one intending the location to be permanent. For example, an officer goes out with a scouting party against Indians, and finding it necessary to wait for reinforcements and supplies, goes into camp at a point where he can get wood, water and grass, and which is also reasonably accessible, until he can hear from Head Quarters. The season for campaigning being practically over, he gets an order to make himself comfortable for the winter. So the whole command is set to work making huts, and in a short time a winter cantonment is formed. The chances are that a good strategic position has been pitched upon by a kind of natural selection. Our supposed expedition has been compelled by force of circumstances to locate near some mountain pass, the crossing of some river, the intersection of roads or trails, and as near to the enemy as it can get with its trains. But as the commanding officer only intended to select a place for a temporary encampment, he made no attempt to select a suitable site for a permanent post; he gave no thought to questions of defense, drainage, extension or beauty. The next spring, however, he receives an order from Department Head Quarters to build a post, and is officially informed that he will be allowed $10,000 out of the Department allotment of barracks and quarters, and that the quarters must be erected by the labor of the troops, etc. Nine officers out of ten on receiving such an order will say, well it won[']t do to throw away all our hard work here; our stables and corral are good enough, and we will put up our permanent barracks and quarters in front of our winter huts and use them for kitchens or out-houses.
>>
>> So, out of a dirty, muddy chrysalis of a winter camp, a poor little butterfly post is started. In another year, another munificent allowance of $10,000 is received or promised out of the barracks and quarters fund, and the cheap adornments of kalsomine [calcimine], whitewash and a little paint is indulged in.
>
> Source: Anderson, "Army Posts, Barracks and Quarters," 421–422.

the government purchased it for a military fort. The confluence of the North Platte and Laramie Rivers had been a site for a trading post since 1834, when Fort William, a wooden structure with high, palisaded walls, was established by Sublette & Campbell, a fur-trading company (fig. 1-2).

Fig. 1-2. Fort William, painted by A. J. Miller, 1837. Fort Laramie was the site of several trading posts before the army acquired it. Fort William, built in 1834 with a high stockaded wall and blockhouse, adheres more to the common conception of an army fort than the fort that was eventually built. Courtesy the National Park Service.

Pierre Chouteau, a rival fur trader, acquired that fort and replaced it in 1842 with Fort John, an adobe, high-walled fort, also known as Fort Laramie (fig. 1-3). In 1849 an army officer described the fort this way: "A square of about 40 yards of adobe wall of 12 ft. height. On the east side are quarters of two stories with a piazza. On the opposite side, the main gate, lookout, flagstaff, etc. with shops, store houses etc. to the height of the wall on their right and left, and on each of the other two sides, ranges of quarters of one story—all opening on a small parade in the centre."[18] By the time the army considered buying the fort, it was not only a viable trading post, but had also become a landmark on the emigrant trail.

As an army post, Fort Laramie was one of three forts established along the Oregon-California Trail in 1849; the other two were Fort Kearny, Nebraska, and Fort Hall, Idaho. A congressional act of May 19, 1846, called for "establishing military stations on the route to Oregon," and these three forts were the result. About a month after the government purchased Fort

26 ■ OUTPOST: FORTS LARAMIE AND BRIDGER, 1849–1869

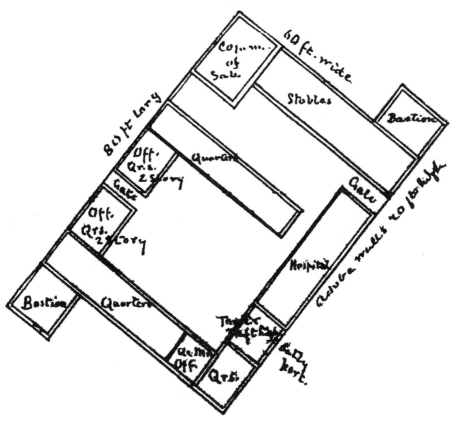

Fig. 1-3. Fort John, plan of fort in 1849, undated drawing. In 1842 rival traders replaced Fort William with adobe Fort John, shown here in plan. From Medical History of the Post [Fort Laramie].

Laramie in 1849, Deputy Quartermaster General Aeneas Mackay inspected the post, traveling 337 miles from Fort Kearny in seventeen days. Fort Laramie was at an elevation of 4,200 feet, located among the rolling grassy hills of the high Plains. High winds and extreme temperatures characterized the site. Mackay's report on Fort Laramie delineated its natural resources, admitting that he had approached Fort Laramie with a negative attitude but that the labors of 1st Lt. Daniel Woodbury of the Corps of Engineers and Maj. Winslow F. Sanderson, the commanding officer of the post, had made evident the advantages of the site. "The rocks—the pines—the timber—the lime—the hay sprang up about them, and Laramie gave promise soon to make herself as desirable a spot as perhaps could be found anywhere so remote from those parts which we are in the habit of thinking only of furnishing us with comfort and happiness."[19] These officers were overly optimistic. They found pine for lumber 12 miles from the site, and driftwood for fuel in the beds of the Platte and Laramie rivers, but both of those sources would soon be exhausted, and scarcity of wood led future residents of the post to consider alternative building materials.

Even under the army's stewardship, Fort Laramie continued to serve as a trading post that furnished provisions to emigrants and traded with Indians through the 1850s and 1860s. As relations with the Indians became increasingly volatile, Laramie served as a base of operations for the army in its expeditions aimed at subduing enemy forces. Fort Laramie was also the site of gatherings that had peace as their objective. The Treaty of Fort Laramie resulted from the assembly of ten thousand Plains Indians at the fort in September 1851. Army officials soon realized that adjacent grasslands were not abundant enough to support their horses, and moved the council 36 miles south to a better site. In 1867 the Peace Commission held a council with the Crows at Fort Laramie. The actual meeting was held in a frame quartermaster's storehouse, converted to hold 250–300 attendees. No treaty resulted from this meeting.[20] The next year, Fort Laramie gave its name to the treaty with Red Cloud that created the Great Sioux Reservation.

Fort Bridger was another trading post on the emigrant trail (fig. 1-4). Established by Jim Bridger, a frontiersman of growing fame, the post had wooden, palisaded walls and a secure appearance. The territory of Utah, created in 1850 with Brigham Young as its governor, encompassed a portion of southwest Wyoming, including the fort. Conflict between Bridger and Young over administration of the post led Young's Mormons to raid the fort in 1853. Although they did not capture Bridger, the hostilities were such that Bridger and his partner, Louis Vasquez, were convinced to sell their post to the Mormons. Rebuilding the post with stone walls, the Mormons continued to use the fort as a supply station for emigrants, although they particularly encouraged emigrants to Salt Lake City, recommending that those headed for California take other routes.[21] In 1857 the U.S. government, grown increasingly wary of Brigham Young, sent the army to remind him of his subordinate status. In the face of military incursions in the Green River area, the Mormons abandoned Fort Bridger, burning it as they retreated, and the army occupied the fort. Troops spent the winter in tents at the site, and the next June the army gave the post the official designation of "Fort Bridger." There seems to have been little debate as to the suitability of the site, as the presence of the army there served to remind the Mormons of the proximity of the government, and it was well located on the overland trail. During the Civil War, Col. Patrick Connor of the California Volunteers occupied Fort Bridger, then moved on to build Fort Douglas on the edge of Salt Lake City.

Fort Bridger was the southernmost point on the Oregon Trail in Wyoming before the trail turned sharply to the north. Later cutoffs had

Fig. 1-4. Fort Bridger, conjectural drawing by William Henry Jackson, 1930. This depiction of the trading post of Fort Bridger shows a stockaded fence and a separate corral. Courtesy Library of Congress.

determined a more direct route to Oregon, but Fort Bridger remained on the heavily traveled route to California and Utah. Located about 50 miles north of the Uinta Mountains, the site had ample wood. A braided stream called Blacks Fork crossed the fort. The location, at an elevation of 6,600 feet, was frequently cited for its healthfulness, and as early as 1876 the site was mentioned as a resort: "Fort Bridger is situated in a beautiful and fertile valley, which is well watered. It makes a very pleasant summer resort and many eastern people are now stopping there, amusing themselves by hunting and fishing."[22]

The recurring issue with Fort Bridger was how big to make the reservation. Every western fort occupied a military reservation—land that had been reserved from public distribution. In July 1859 President Buchanan declared that Fort Bridger's military reservation would stretch 20 miles east and west and 25 miles north and south, intended to give the fort a wide buffer from its Mormon neighbors. This 500-square-mile area soon proved a disadvantage, because without nearby settlers, the post had either to buy all its goods from the post trader, William Carter, or have them shipped from Salt Lake City. In 1866 Col. James F. Rusling of the Quartermaster's Department concluded that both alternatives were too expensive and argued for a reduction of the reservation. As Capt. Edward B. Grimes reported at about the same time, "If opened to settlement a large community would soon be formed which would furnish grain, hay, beef and vegetables to the Govt. at cheap rates." Nearby settlements were not always viewed so positively; they were also seen as vice-ridden communities established to prey upon soldiers. At Fort Bridger, the advantages of competitive producers outweighed such concerns, and the reservation was reduced to 4 square miles in 1871.[23]

Fortifications

Once the army decided to build forts at these former trading posts, the first impulse was to construct a fortification, reflecting army defensive strategy. But fortified posts turned out to be the exception, not the rule, in the West, reversing the usual trend in fort design. Traditionally, forts were designed to withstand attack both passively, by impregnable siting and massive stone construction, and actively, by providing secure sites from which defenders could fire at attackers. When facing an enemy without heavy firepower, such as Indians, the army was satisfied with stockades, or tall walls of vertical posts or pickets. Corner blockhouses—fortified buildings from which defenders could fire along the walls—served the purpose of bastions in masonry forts. Inside the fort, buildings were typically aligned along the perimeter, so that the rear walls functioned as part of the stockade. Forts on the Missouri River, which secured the navigable waterway and also protected emigrants, followed this model.

Other stockaded forts included those along the trail from Fort Laramie to Bozeman, located to protect the prospectors' route to the Montana gold fields just after the Civil War. Col. Henry B. Carrington led a regiment up the road in 1866, garrisoning Fort Reno, and building Fort Phil Kearny and Fort C. F. Smith in the face of constant harassment from resentful Sioux. Like all of these forts, Fort Phil Kearny, Carrington's headquarters, had a stockade (fig. 1-5). Measuring 600 by 800 feet, it was composed "of pine trunks, length 11 feet, tamped 3 feet in gravel, side-hewn." But Col. William B. Hazen inspected the post during construction and questioned the priorities. Government property suffered without a storehouse in which to place it, and it was unlikely that barracks and quarters would be completed before winter, but Carrington "has built a strong stockade, the best I ever saw, except those built by the Hudson Bay Company." Hazen attributed Carrington's priorities to the presence of women: "Had there been no officers' wives with the command . . . I believe the labor on the stockade (some two months) would have been first applied to store-houses

Fig. 1-5. Fort Phil Kearny, 1867. Col. Henry B. Carrington built Fort Phil Kearny on the Bozeman Trail to assert dominion over the route to the gold fields. Because of anticipated conflicts with Indians, he had the soldiers build a stockade. From *Journal of the Military Service Institution of the United States* 1, no. 6 (1881): frontispiece.

and quarters for the men, which would be now, probably, enclosed." The Sioux never attacked the fort, but they did attack a party of men sent out to obtain wood. In the subsequent investigation, Carrington was criticized for devoting too much of his energies to defensive construction and not enough to training.[24]

Although stockade walls were not universal, they formed the common image of a western fort. Even in the late nineteenth century, visitors expected stockades and expressed disappointment when they were absent: "A plains fort is no fort at all; it is simply a collection of houses and buildings set down on the prairie or on the crest of some high bluff, with no bastions, walls, stockade, nor defence of any kind, and might better be termed a small settlement than a fort."[25] Even today, modern Americans raised on Hollywood movies and television shows such as *Rin Tin Tin* and *F-Troop* expect every western fort to have a high, palisaded stockade with blockhouses at the corners and a prominent gate that is closed just in the nick of time against savage pursuers.

In reality, Fort Laramie's various phases of fortification were more typical, reflecting erratic attempts to fortify the post. As a trading post, Fort Laramie consisted of high, windowless, adobe walls, presenting a fortress-like appearance. By the time the army acquired it in 1849, the old fort was crumbling. Lt. Daniel Woodbury of the Corps of Engineers was responsible for the purchase of the trading post and the establishment of the fort (fig.

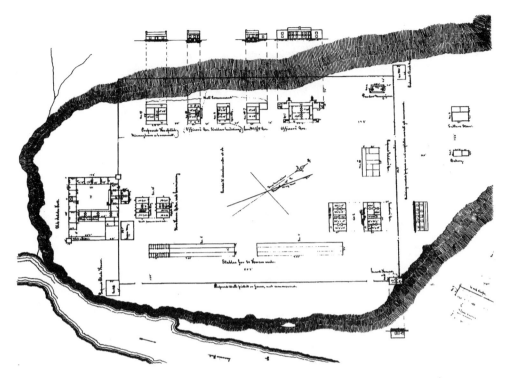

Fig. 1-6. Fort Laramie, 1851. Although Lt. Daniel Woodbury's plan displayed a stockade wall and orderly buildings within, neither the stockade nor most of the buildings were ever built. Once Woodbury left the fort, no additional buildings were constructed to his plan. Courtesy National Archives.

1-6). Adjacent to the 123-by-168-foot trading post, Woodbury planned a 550-by-650-foot enclosure, which he described as follows: "The enclosure may be made by a fence 9 feet high or by a rubble wall of the same height laid in mortar, at the discretion of the commanding officer. If a fence, the posts should be about 10 feet apart, average 12 inches in diameter and enter four feet into the ground. The boards should be nailed on upright, close together, to three horizontal ribbons, on pieces 4 inches wide, 1-1/2 inch thick, and pointed at top." The cost of this wall, which Woodbury estimated at $12,000, was prohibitive. At opposite corners Woodbury proposed 30-by-40-foot blockhouses. The wooden second story of these blockhouses overhung the stone first story by more than one foot. The wood frame was infilled with adobe, and the walls were covered inside with 1-inch boards up to a height of 6 feet, both serving as protection from bullets.[26] At a third corner, Woodbury oversaw construction of a guard house, about 17 by 31 feet, with a stone basement and wood-frame second story.

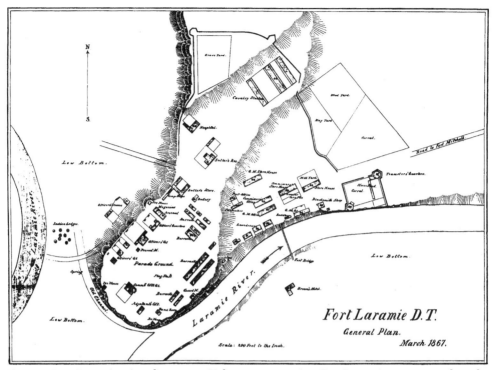

Fig. 1-7. Fort Laramie, site plan, 1867. Volunteers manning Fort Laramie constructed earthworks on the north side of the fort in 1865. The next year, regular troops built an adobe redoubt on the east; at the time of this plan it functioned as teamsters' quarters. Courtesy National Archives.

The Engineer Corps's involvement in the construction of this post was unusual. Traditionally, construction of posts on the inland frontier was the responsibility of the quartermaster general, while the engineers controlled seacoast fortifications. Woodbury's actions at Fort Laramie were in association with the establishment of three posts along the Oregon Trail and the fact that they were to be fortified. Chief Engineer Col. Joseph G. Totten withdrew Woodbury and his assistant, 2nd Lt. Andrew J. Donelson, in August 1850, once their efforts were expended "solely upon the construction of quarters, barracks, kitchens, stables etc. etc., objects pertaining to the accommodation of the garrison alone." They turned the work over to the Quartermaster Department, with the appropriation exhausted.[27] No wall was ever built.

Cost was just one reason why stockaded walls were not routinely built in the West. In addition, in areas where timber was scarce, a stockade would have consumed an inordinate amount of wood. Further, a stockade, or any fortification at all, was intended to enclose the whole fort and all of its troops. With the introduction of cavalry, forts grew to accommodate mounted

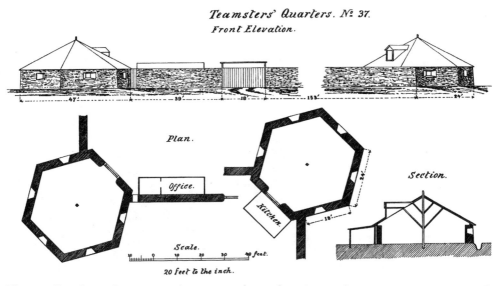

Fig. 1-8. Fort Laramie, teamsters' quarters, plans, elevation, and sections, 1867. Hexagonal adobe blockhouses had rifle embrasures on their outside walls and a dormer window on the side that faced into the corral. Courtesy National Archives.

troops, which occupied a much larger space than infantry. Encircling this larger parade ground would have been a much more difficult task. Moreover, the army in the West soon realized that fortifications were unnecessary since Indians rarely attacked fixed positions. They were much more likely to harass details or to run off grazing animals. Other reasons for not having stockades include a show of courage, as related by Col. Philip Regis de Trobriand, commander of the Department of Dakota: "General Stanley believes that in dealing with such a contemptible enemy as the Indian, it is better for troop morale to depend on vigilance and breechloaders for protection than to hide behind palisades. I think the general is quite right."[28]

But without a stockade, Fort Laramie was still seen as vulnerable to attack. In 1865, in an increasingly tense atmosphere created by Col. John Chivington's massacre of Indian women and children at Sand Creek, Colorado, another effort was made to fortify Fort Laramie. Volunteer forces manning the post constructed three earthwork batteries on high land north of the fort (fig. 1-7). The next year, regular forces commanded by Maj. James Van Voast built a fortified adobe redoubt that doubled as a corral. Rather than an attempt to enclose the whole fort, this was a small defensive structure located northeast of the parade ground, along the river, designed to serve as a stronghold. The 8-foot-high adobe wall enclosed an area of about 2 acres. At two opposite corners stood hexagonal blockhouses, 24 feet on a side (fig. 1-8). Never used defensively, the blockhouses served as teamsters'

Fig. 1-9. Fort Laramie, 1867. Corresponding closely to the 1867 plan, this view shows the general sprawl of shops and storehouses off to the right, with the adobe redoubt on the far right. On the artist's side of the river are "Brown's Mess House," a bar and restaurant for soldiers, and on the far left, Indian tepees. Courtesy American Heritage Center, University of Wyoming.

quarters, and the redoubt functioned as the quartermaster's corral well into the 1870s (fig. 1-9).[29]

When he toured western posts in 1866, General Sherman mentioned, "It seems to me as a general rule all these posts are built too scattered, so as to require too many men for defense. They should have two or three block houses so placed that a few men in each would cover all the ground by their fire. This would enable all the garrison to go out." The blockhouses would be manned by civilian workers: "There are now and will probably always be many citizens employed here who could defend themselves and the post in an emergency, leaving the garrison free to go out." Less than a week later, Brig. Gen. Philip St. George Cooke, commanding Department of the Platte, turned Sherman's suggestions into this order: "Commanders building or reconstructing Posts in this Department will so plan them that, by blockhouses—of logs, stone, or adobes—and otherwise, a few men shall be sufficient for defence; and thus enable the greater part of the garrison to act far beyond; their only objects being to protect migration, the mail, the telegraph, commerce, travel, and settlements."[30] The blockhouses/teamsters' quarters at Fort Laramie resulted from this order. The motive was not to turn forts into defensive structures; it was clear by this point that the army was on the offensive against the Indians. Still, troops needed a base, and

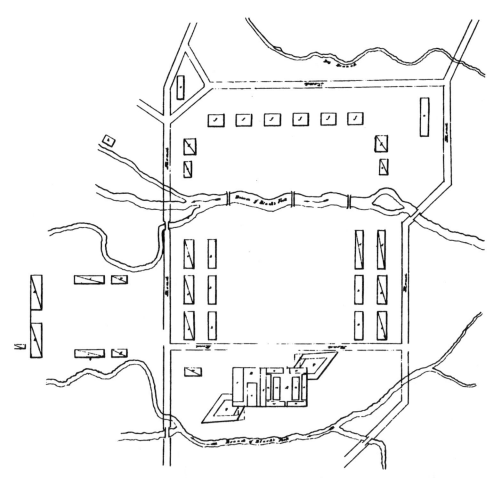

Fig. 1-10. Fort Bridger, 1859. When the army occupied Fort Bridger in 1858, it consisted of the stone fort and corral at the bottom of this plan. The army added lunettes. Most of the buildings indicated here had not yet been built, but six officers' quarters and five barracks had been constructed on the parade ground. Courtesy National Archives.

"a few men" strategically placed in blockhouses could defend the station while larger bodies of troops were aggressively "protecting," "far beyond."

Fort Bridger also had fortifications, at least as long as the threat of immediate hostilities persisted (fig. 1-10). In this case, the hostile forces were Mormons, and the threat lasted only briefly. After seizing the post in November 1857 in the wake of the Mormons' departure, Col. Albert S. Johnston took the stone fort, measuring 100 by 100 feet, and adjacent corral, measuring 80 by 100 feet, and added lunettes, or fortified projections, at two opposite corners. With this, he had a small stronghold and a place to store supplies. The troops, numbering 2,500 along with civilian employees, wintered in tents.[31] The next summer they built a larger, unfortified post; the original stone fort was devoted to storehouses.

Layout

The army forts in the West without walls or blockhouses kept on being built in the same configuration as if they were fortified. Thus, a parade ground formed the heart of the fort, with the primary buildings arrayed around it, facing inward. The parade ground was used for the assembly of troops, which typically occurred three times a day for roll calls and twice a day for drills, although the latter were suspended when the post was shorthanded. The arrangement around a parade ground was understandable with a fort like Laramie that was originally planned to be enclosed, but Fort Bridger adopted the same convention. Most forts arrayed the officers' quarters on one side, with barracks opposite; many located the guard house on the third side, with headquarters opposite. Behind the officers' quarters were kitchens and outhouses; behind the barracks were messhouses, latrines, and laundresses' quarters; and off the parade ground were storehouses, corrals, hospital, and all the other buildings associated with the post. The inward focus of a post without walls meant that no imposing front faced visitors, but rather the outbuildings made the first impression. One astute observer of army posts, Lt. Col. Thomas M. Anderson, pointed out that "nearly all our posts are built with the barracks and quarters facing inward on an oblong parade, hence the visitor must enter by a back door." He recognized the traditional origins of this arrangement, but questioned why contemporary forts were being built in the same manner. He recommended the model of a city square, in which all the buildings face outwards, but this plan gained no followers.[32]

Another model for fort layout was the temporary encampment, in which rapid egress was key. But many exits means many entrances, and hence a certain permeability. Few unfortified forts had gateways or recognized entrances; they could be entered at any point. The porousness that resulted is illustrated by an incident that occurred at Fort Laramie in 1864 when a patrol searching for Indians returned to the fort. The soldiers unsaddled their horses and let them roll on the parade ground, while the officer in charge reported that there were no Indians in the vicinity. At that moment, about thirty Indians swept into the fort and stole the horses off the parade ground. They were never caught. Similarly, wild animals could pass through the parade ground unimpeded. In a story titled "A Raid on Fort Laramie," the newspaper reported how "a huge elk . . . ran out of the underbrush in the Platte bottom and charged directly through the parade ground. The antlered monster was pursued by a pack of dogs of all sizes." The elk killed several dogs before escaping, despite the pursuit of a dozen men and the firing of "several hundred cartridges."[33]

Fig. 1-11. Fort Laramie, officers' quarters, 1885. Although planned as a cohesive unit, Fort Laramie's officers' row had a mixture of buildings constructed over several decades, resulting in a noticeable lack of uniformity. Courtesy National Archives.

Topography inevitably influenced the layout of a post. A bend in the Laramie River bordered a bluff—a natural location for a fort. Fort Laramie's parade ground surrounded by barracks and quarters was centered on this high ground, which in this instance limited the fort's size and shape. Secondary buildings, such as storehouses and civilian workshops, occupied lower ground, off the bluff.

Fort Laramie also exemplified a post that grew and changed in accretions. As new buildings replaced old ones, their sites often changed, disrupting the original symmetry of the plan, which had officers' quarters on two sides, facing barracks and stables on the other two sides (fig. 1-11). Woodbury's plan indicated stables on the southeast side of the parade ground, but barracks were built there instead; in 1885 an administrative building appeared there as well. While the west side of the parade ground remained devoted to officers' quarters, these were built over four decades and maintained no consistency in form or materials: one and two stories; wood, adobe, and concrete walls; single, double, and multiple units; gable and mansard roofs.

Fort Bridger followed the standard convention of officers' quarters along one side of a large parade ground, barracks on the two adjacent sides, and storehouse buildings on the fourth side. The parade ground measured about 375 by 645 feet, but instead of being an open area suitable for the parade of mounted troops, it was traversed by a stream. The branch of Black's Fork that crossed the parade ground provided fresh water and trout but detracted from the military atmosphere. The overland stage road also crossed the military reservation, running behind barracks and storehouses on the north side of the parade ground. Aside from these intrusions, Fort Bridger's plan, which never included a stockade, was much like Fort Laramie's, which was predicated on a stockade.

The products, then, of government policy, military tradition, and exigencies of the site, Forts Laramie and Bridger both exhibited somewhat erratic appearances. Preexisting trading posts dominated the sites and initially provided the most fortified areas. The forts had no main entrance, but were essentially porous. Both forts focused on open, rectangular parade grounds, but a clutter of outbuildings spilled out in every direction. Despite the informality of their layouts, the forts were far more orderly than the surrounding countryside or Indian encampments. To emigrants and visitors, the forts had a civilized appearance and were easily recognized as forts. This tension between order and informality, between the ideals of what a fort should be and the realities that the site and circumstances imposed, is reflected even more clearly in the individual buildings.

2

HUTS AND ADOBE BUILDINGS SADLY IN NEED OF REPAIR

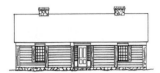

Architecture

The buildings constructed at Forts Laramie and Bridger before the Civil War embody contradictions in intention and execution (fig. 2-1). The army had no explicit standards for its architecture, only a vague sense of tradition. The planned buildings tended to reflect military precedents and eastern civilian norms, as can be seen in plans, materials, and ornament. But the realities of site and situation often rendered these ideal examples impractical. Initially, post commanders turned to local customs and materials as a means of dealing with a landscape of unfamiliar resources. The employment of adobe, whose use had migrated from the Southwest, and paneled log construction, coming from Canada, indicated expedient approaches, adoption of available materials that had been proven successful. Similarly, the first officers' quarters constructed at a post were often linear arrangements of rooms without halls, more typical of Spanish and French colonial buildings than English, signaling an adoption of regional norms.

The army's preferred architecture consisted of symmetrical, rectilineal buildings with gable or hip roofs, passageways and porches, and a commanding presence. The preference for classical ideals of order and symmetry had influenced the design of houses of the East Coast gentry beginning in the eighteenth century. The term for the double-pile, center-passage plan, "Georgian," reflects its popularity among Anglo-Americans through its reference to eighteenth-century kings of England. While historian Richard Bushman points to the association of the center-passage plan with gentility,

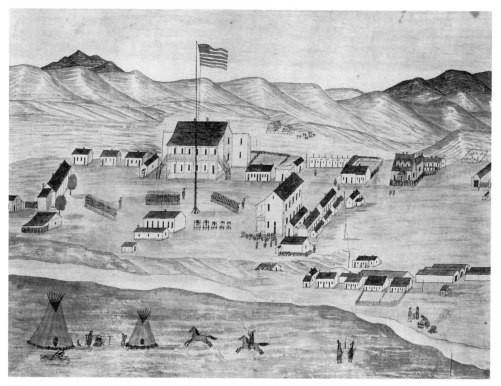

Fig. 2-1. Fort Laramie, drawing by Charles Moellman, ca. 1863. Trumpeter Moellman evidently thought that the officers' quarters behind the flag was the most imposing building, but he also portrayed a variety of activities at the fort. The dark-colored building to the right is the post trader's house; left of it is the post trader's store, with a flat-roofed wing to the left of company shops. The laundresses' quarters are in the lower right, near the laundress working at the riverbank. Most of these buildings appear on the 1867 site plan (fig. 1-7). Courtesy American Heritage Center, University of Wyoming.

he also notes that it soon filtered down to middle classes and became common throughout the eastern United States as a sign of bourgeois status. The popularity of the form extended to initial attempts at standardized designs for the army, published in 1860 and 1872, which employed this configuration. Plans for double officers' quarters, however, often placed the passages in the outside bays, not the center, departing from the prototypical Georgian plan. Still, neoclassical notions of order and symmetry survive, symmetry being particularly useful for producing equal-sized quarters for officers of the same rank. The appearance of the passage was also significant; although it did not provide usable square footage in terms of living space, it was an important feature for genteel living, separating visitors and servants from private family space. Although these plans may have differed superficially, they represented a common style, or way of conceiving a building.[1]

The first wave of buildings at Forts Laramie and Bridger did not always exhibit this orderliness. This contradiction between real and ideal results from the various constraints imposed upon army officials. The labor of both design and construction fell largely on the officers and soldiers at the post—men with little or no training in these fields. The primary guidance issued by the army was in the form of regulations specifying the amount of space to which each man at the post was entitled. In an effort to manage inadequate budgets, post quartermasters used local materials, including green lumber and poorly maintained adobe. The army's insufficient funding—reflecting public attitudes toward the army—was readily apparent in the forts that it built.

Standard Designs in the 1860s

Standard designs, which could be used from post to post, offered some advantages to the producers of fort architecture. Despite several attempts in the 1860s to implement uniformity, however, such plans failed to survive the exigencies of distance and local autonomy. The most intriguing effort to standardize military architecture in the West—and certainly the most comprehensive—was the 465-page volume produced by Don Carlos Buell of the Adjutant General's Office. Buell was an 1841 graduate of West Point who fought in the Mexican War. He transferred to the Adjutant General's Office shortly afterward and served at several posts. In 1859 he was assigned to the War Department in Washington, where he apparently supervised production of the volume. (Buell's name does not appear on the work, and the primary attribution rests with John Shaw Billings of the surgeon-general's office, who lauded these discarded plans in an 1870 survey of posts.) At the outbreak of the Civil War Buell was given command of the Army of the Ohio, but he resigned in disgrace when questions were raised about his behavior during battle and he was accused of being a southern sympathizer. His comprehensive regulations remain something of a mystery, as his particular role in their production, as well as the reasons for their rejection, is unknown.

Published in 1861, Buell's work provided plans for fourteen types of buildings. Although the illustrations showed a mortise-and-tenon framing system with board-and-batten siding, notes explained that these buildings could be constructed of any material. Specifications for construction in frame, stone, brick, earth, or logs for each of the fourteen buildings accompanied the drawings. Post quartermasters were to decide the best material and estimate its cost, using the quantities provided in the specifications.

All of Buell's designs featured wide passageways and porches, providing intermediate spaces for a mixture of public and private uses. The designs seemed to be planned for a warm climate, with tall windows, extensive

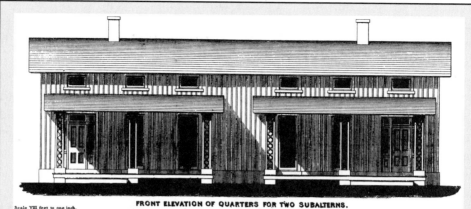

Officers' quarters, elevation, 1860. From *Regulations concerning Barracks and Quarters for the Army of the United States, 1860*, pl. II.

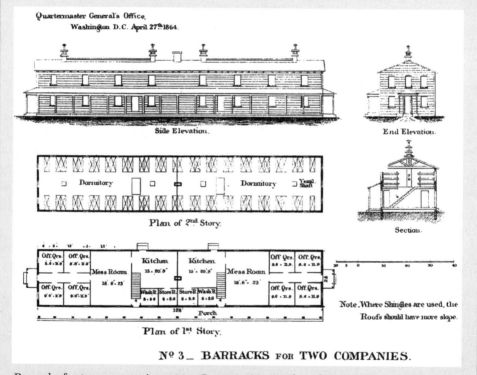

Barracks for two companies, 1864. Courtesy National Archives.

cross-ventilation, and detached kitchens for the captains' and lieutenants' quarters. He stretched the allowances for rooms, providing a lit attic space with undefined use for most officers' quarters. The one-room-plus-kitchen plan that he provided for a lieutenant was augmented by a generous 10-foot-

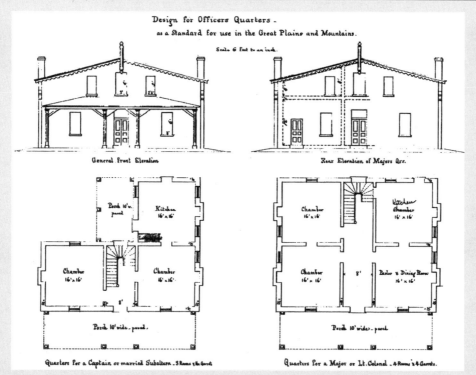

Officers' quarters, plans, elevations, 1867. Courtesy National Archives.

wide hall and 10-foot-wide porches. The main room was also expansive, measuring 17 feet 4 inches by 18 feet.

The Quartermaster General's Office made at least one attempt at standardized architecture during the Civil War. On April 24, 1864, it published plans for officers' quarters, barracks, hospital, and stables. The officers' quarters consisted of a two-story block with eight rooms on each floor, separated by both longitudinal and transverse corridors. The building could have housed sixteen lieutenants, eight captains, or, more realistically, however many officers it had to. Constructed of wood frame, the design was very plain. The barracks for two companies was of more interest, as Quartermaster Gen. Montgomery Meigs returned to this plan later. In 1864 he proposed a two-story building with two open dormitories on the second floor where men slept in three-tiered double bunks. On the first floor, the mess room and kitchen were the largest rooms, with smaller store room, wash room, and four rooms for officers. Despite the crowded conditions—108 men sleeping in a 63-by-23-foot room with a floor-to-ceiling height of 8 feet, yielding a little more than 100 cubic feet of air space per man—Meigs showed some concern for air quality by the introduction of a ventilating stove, which drew air in from outdoors underneath the stove and drew stale, cold air from the room up through the chimney box.

> The allocation of $1 million in construction funds to the Military Division of the Missouri in 1866 set off a mad scramble for practical designs that could be constructed at reasonable prices—a situation that standard plans might assist. On May 25, 1867, the chief quartermaster of the Department of the Platte, William Myers, sent "plans of proposed Officers' Quarters, barracks, and storehouses, to be used at Posts in the Department of the Platte" to his superior officer, James L. Donaldson, chief quartermaster of the Military Division of the Missouri. Three days later, Col. Langdon C. Easton, chief quartermaster of the Department of the Missouri, sent copies of his department's standard plans to Myers. Surviving plans labeled "Design for Officers Quarters as a Standard for Use in the Great Plains and Mountains" could be either those of Myers or Easton. The drawings show quarters with three, four, and five rooms, depending on rank. All have center halls in a symmetrical arrangement, a common American house plan. The shallow-pitched, front-gable roofs are decidedly uncommon, as usually the roof ridge is parallel to the longest side of the building. Easton claimed that Maj. Gen. Winfield S. Hancock, commanding the Department of the Missouri, "has directed that these plans be followed in building new posts or in additions to posts where desirable uniformity would not thereby be destroyed." If these drawings were used, no known examples survive.
>
> Sources: *Regulations concerning Barracks and Quarters for the Army of the United States, 1860.* Cullum, *Biographical Register of the Officers and Graduates of the U.S. Military Academy* 2: 95–96. U.S. War Department, Surgeon-General's Office, Circular No. 4, *Report on Barracks and Hospitals with Descriptions of Military Posts*, xxv. Donaldson to Myers, 1 June 1867, Box C–D, and Easton to Myers, 28 May 1867, Box E–F, RG 393, Part I, Entry 3887, NAB. Grashof, "Standardized Plans, 1866–1940, 1: 10–11, 2:OQ-5, OQ-6, OQ-7.

As Built

While army officials may have had lofty goals and well-ordered plans for the architecture at a post, it is important to examine what was actually built and how it fared over time. Only the sturdiest and most substantial buildings have survived into the present, while large buildings such as barracks, which had few options for reuse, rarely lasted. The quickly built shops and dwellings for civilian employees, located off the parade ground and never a military priority, disappeared the most readily. Although sometimes photographed in general scenes of the fort, these utilitarian buildings usually received little documentation in the graphic or written record.

One exception to this incomplete record exists in an extraordinary set of documents for Fort Laramie. In 1867 an accomplished draftsman made architectural drawings of every building at the post. When compared to a written description of the buildings in a list that accompanied the drawings and to the post return of March 1867 (the month and year given on the building list and site plan), a clear picture emerges of the fort at that time. The poor quality of the buildings is a constant refrain. There was also a mix of materials used: some structures were wood-frame, but many were adobe, reflecting an expeditious solution to the unavailability of lumber. Flexibility in function was also required, as post quartermasters scrambled to find space when needs dictated.

In 1867 a tense military situation stemming from the incident at Fort Phil Kearny caused the army to assign eight companies to Fort Laramie, swelling the population of the post. The March post return lists 13 officers, 459 enlisted men, 9 Indian scouts, and 300 civilian employees; women and children were not included. Forty-seven army buildings provided housing and work spaces for these people; the condition of nearly half of the buildings was given as "bad," "very bad," "unserviceable," or "useless." Twenty of the buildings were adobe, and nineteen were "rough boards, upright and battened," none of which survive.[2] The only survivals of this period are one set of weatherboarded officers' quarters, a stone guard house, and a stone magazine. (The post trader's store also survives, but as it was not government property, it was not included in this inventory.) Although the drawings provide an important record, they are, if anything, too neat, imparting a sense of regularity and stability that was absent in many of the buildings themselves. Many of the buildings were old and rickety, ephemeral in nature and poorly built and maintained.

The adobe trading post that preceded the army's possession of the site in 1849 remained standing until 1862. In the first year and a half of possession, the army built four temporary adobe buildings and three temporary "slab" buildings, all off the parade ground, for workshops, sawmill, and quarters for civilians. On the parade, soldiers built more formal structures intended to last: three framed buildings filled with adobe, resting on stone foundations, as officers' quarters, barracks, and guard house. The powder magazine was stone, covered with earth.

One of these early buildings, proposed by 1st Lt. Daniel Woodbury of the Corps of Engineers, survives (fig. 2-2). Quarters intended for four officers consisted of a two-story building with exterior staircases and flanking, two-story kitchen wings. If the building was used as intended—and, in fact, four officers were stationed at Fort Laramie in 1852—each officer was assigned two rooms, a hall, and a kitchen.[3] But as the post grew, the uses of Bedlam, as the building was called, also changed. (Referring to a bachelor officers' quarters as a wild place, "Bedlam" derived from the name given to a lunatic asylum in London.) During the Civil War, Lt. Col. William O.

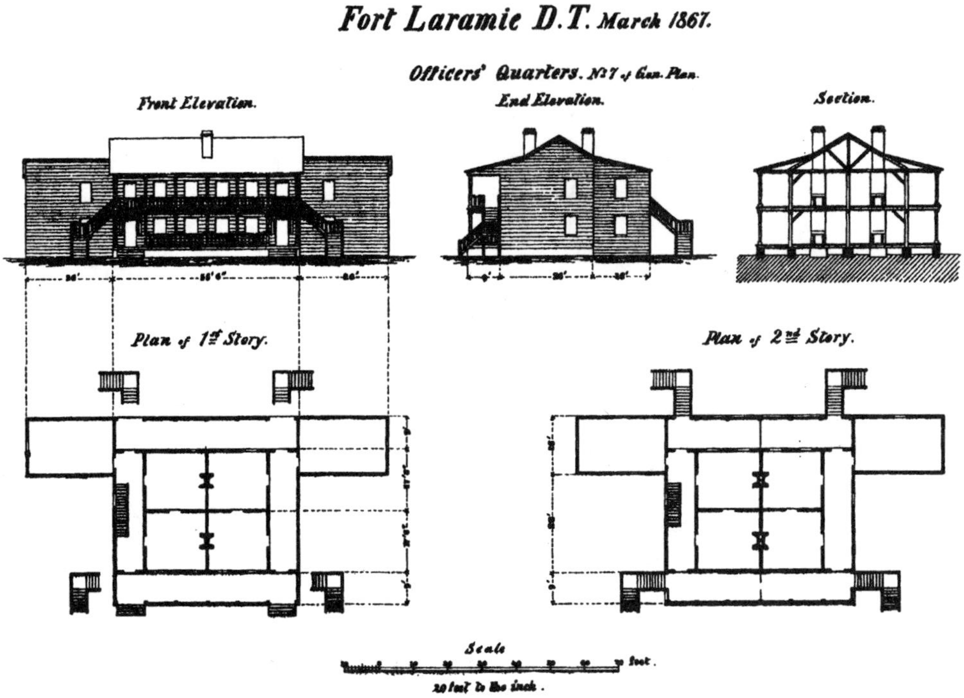

Fig. 2-2. Fort Laramie, officers' quarters No. 7, plans, elevations, and section, 1867. Officers' quarters constructed in 1849 were adapted to several configurations of living arrangements. Courtesy National Archives.

Collins of the 11th Ohio Volunteer Cavalry occupied the west half of Bedlam, using the second floor as quarters for his family. One room downstairs served as post headquarters, while the other was a dining room shared by all the officers. Probably at this time (and certainly by 1866), an interior stairway was added in the west hall to facilitate this usage.

Porches front and rear, on both first and second stories, permitted additional internal patterns of traffic flow. They were also featured prominently in a novel about Fort Laramie, written by an army officer named Charles King. *"Laramie;" or The Queen of Bedlam*, published in 1889, was a romance in which the ladies at the fort—officers' wives and their friends and sisters—played a major role. The access granted by exterior porches enhanced the story's mystery; with a swish of silk in the night an unknown being swept through the officers' quarters. The porches also accounted for the building's tropical appearance, according to one view. Frenchman Louis Simonin wrote, "With its two-story 'veranda,' or outer gallery, one would take it for a hotel in Panama or Central America."[4] Porches enhanced most residential buildings at western posts. Their functions were numerous: providing protection to windows from hailstorms, as suggested by Fort Laramie's post surgeon; giving access to multi-unit dwellings, as is the case here; furnishing shelter from the sun in summer, as was certainly likely at Fort Laramie; and serving as a transitional gathering place. Porches were a prominent aspect of military buildings.

The two-story kitchens attached to Bedlam were judged "not fit for the purpose"; hauling water and wood to the second story was an onerous task for servants.[5] In 1881 workers demolished the kitchens and replaced them with a one-story stone wing in the rear of the building. The new wing housed two kitchens, reflecting Bedlam's new function as double officers' quarters. Like many military buildings, Bedlam's use changed over the decades: from four units of officers' quarters, to half of it devoted to the commanding officer's quarters, to two units of officers' quarters.

This multi-unit officers' quarters was an imposing building. With its weatherboarded exterior, two-story height, porches and gable roof, and a floor plan that included passages, it was a familiar type in mid-nineteenth-century America. Its symmetry and order contrasted with the other officers' quarters present in 1867: three one-story adobe buildings without halls (Nos. 3, 4, and 5). Each of these had two units; each unit consisted of one room with a fireplace, and two unheated rooms in a wooden lean-to.

The various commanding officer's quarters at the fort exhibited a similar dichotomy (fig. 2-3). In 1867 the commanding officer's quarters (No. 2) was a one-story building divided into four linear rooms; two large frame wings in the rear were not heated. The board-and-batten quarters had no hall and no internal connections between the four front rooms. Constructed in 1858, the quarters was described as "serviceable at present but old and rickety." In 1867 a new commanding officer's quarters west of this building was constructed; here, concepts of eastern gentility and symmetry were evident. The building had a center-hall plan, four main rooms, and a kitchen wing. (Curiously, the design was similar to that issued five years later by the quartermaster general as model plans.) The next year, construction began on yet another commanding officer's quarters, east of the first one (fig. 2-4).[6] The wood-framed, adobe-lined building had a similar plan to the previous one. However, the building never served as commanding officer's quarters. The subsequent commanding officer, Lt. Col. Franklin F. Flint, declined to occupy it, without giving a reason. When the building was nearly completed, simple adjustments were made: a partition wall divided the building in half, and two front doors replaced a single one. A second kitchen wing was added, forming a U-shaped plan, and the building served as double quarters for officers.

Only officers lived in quarters that resembled houses; soldiers inhabited barracks. At Fort Laramie, the barracks exhibited a range of materials and plans. The first barracks, like the four-unit officers' quarters, was a two-story building of wood frame, weatherboarded (fig. 2-5). This building, based on

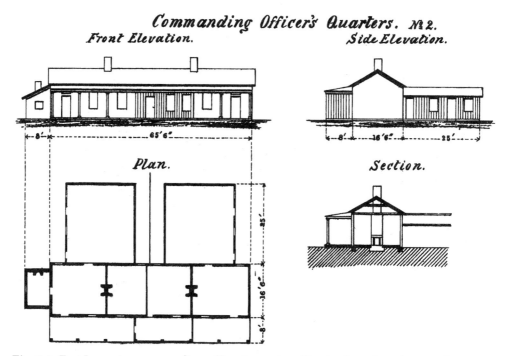

Fig. 2-3. Fort Laramie, commanding officer's quarters No. 2, plan, elevations, and section, 1867. The commanding officer's quarters, built in 1858 but by 1867 "old and rickety," was replaced that year. Courtesy National Archives.

Lieutenant Woodbury's designs, reveals the army's desired accommodations. Intended for one company, the building had a two-level porch across the front. The second story included four rooms measuring 16 by 30 feet, apparently intended as sleeping quarters. Numerous doors on both the first and second levels permitted free entry and egress. But the construction was never completed. By 1856 the building housed two companies, and the post quartermaster complained that it "is yet unfinished, never having been plastered and only partially lathed. The piazzas are not finished, and the roof is in such a bad condition, that it should be at once renewed. In consequence of its unfinished condition this building is very open, and those quartered in it suffer much from cold during the winter."[7] In 1867 its condition was described as "very bad, upper story useless, soon falling down," although it was occupied.

In the fall of 1866 Captain and Assistant Quartermaster George Dandy struggled to provide shelter for eight companies. He had available the two-story barracks described above, as well as two one-story, one-company barracks (Nos. 17 and 18), each measuring 76 by 24 feet, divided into two rooms. Built in 1855 of adobe brick, they were described just twelve years later as "unserviceable." Dandy supervised construction of a new, two-

Fig. 2-4. Fort Laramie, officers' quarters, 1891. Built ca. 1868–1870 for the commanding officer but never occupied by him, this board-and-batten quarters was divided into two units for officers. Courtesy Library of Congress, HABS.

company adobe barracks, 212 feet long (No. 22; fig. 2-6). He experienced difficulty obtaining lumber for roofs and floors, due to Indian harassment at the sawmill, 55 miles west of the post in the Laramie Mountains. Unable to build a second barracks before winter, Dandy proposed to cram four companies into his new building by using three-tier double bunks.[8] Dandy thus put two companies in the 1855 barracks, two in the crumbling 1849 barracks, and four in his new, two-company barracks.

The 1867 drawings also depict lesser residential buildings. At the prescribed rate of one laundress for every nineteen enlisted men, there should have been twenty-four laundresses at the post, but only three buildings of laundresses' quarters, each divided into three units, were recorded. Located near the river, two of these buildings (Nos. 32 and 33) were adobe, offering fireplaces in each of the two rooms in each unit. But the other building (No. 31) was frame with board-and-batten siding and a shed roof; it had two fireplaces for the three units (fig. 2-7). The inspector described the laundresses' quarters as "utterly worthless, recommended to be torn down," but they were not replaced for five years.[9]

Housing 300 civilian employees was obviously a challenge. Of those, 170 were teamsters; another 31 were wagonmasters, wheelwrights, and

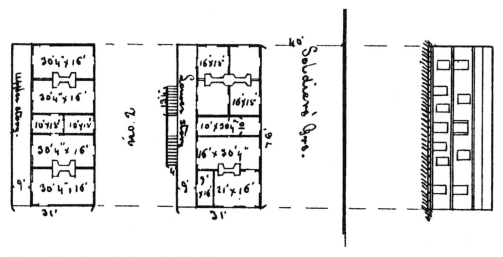

Fig. 2-5. Fort Laramie, barracks, detail of 1851 map. This early drawing of the first barracks at Fort Laramie shows the division of dormitory space on the second floor into four rooms, not the large open space that would prevail in later plans. Seventy-four men occupied this building, with their kitchen and mess hall on the first floor. Courtesy National Archives.

hostlers; 67 apparently worked in construction; 11 served as guides and interpreters; 9 clerked or kept store; and the rest included a master mechanic, a fireman, 9 watchmen, and a miller. Only three buildings were specifically classified as housing for these men. A modest, three-room building labeled "laborers' quarters" (No. 21), constructed in 1849 of frame clad with weatherboards, had deteriorated and was demolished in 1867; a six-unit adobe building (No. 35) was constructed in its stead (fig. 2-8). The "teamsters and mechanics quarters" was the sod corral that doubled as a redoubt (No. 37). In 1873 the commanding officer described the living arrangements of civilian employees: "The mechanics are chiefly quartered in rooms in rear of their respective shops, or in temporary buildings made of slabs covered with condemned [surplus] canvas."[10]

One of only two stone buildings at the post was the guard house, built in 1866 (fig. 2-9). The building was set into the bluff, so the main story faced the parade ground and the basement faced the river. The main story housed the sergeant of the guard in one room with a "rough board bed," a desk, and a few chairs. An adjacent room accommodated members of the guard not standing at post. The basement, which according to the post surgeon had no furniture, heating, or lighting, held prisoners. Guard houses received the attention of the surgeon-general's office, which was most concerned with adequate ventilation.[11] Prisoners confronting the cold, damp, stone floors and walls of the Fort Laramie guard house were probably more concerned with heat. Within a decade, this stone guard

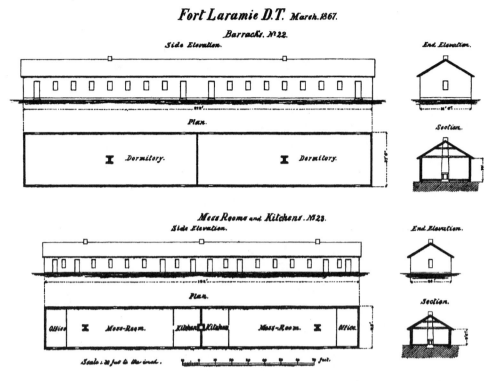

Fig. 2-6. Fort Laramie, barracks No. 22, plans, elevations, and sections, 1867. Built in 1867 of adobe, this barracks was designed to hold two companies, with their respective mess rooms and kitchens in a separate building behind. Courtesy National Archives.

house was replaced with a concrete one that closely followed model plans issued in 1874.

About one-third of Fort Laramie's buildings were not residential in nature. These included offices for the quartermaster, provost marshal, post adjutant, and post office, and shops for the carpenter, wheelwright, painter, saddler, blacksmith, and tailor. These tended to be small, one- or two-room buildings; larger buildings included the stables and storehouses. None of these buildings was fancy, and few earned descriptions of any kind in the written record. But they contributed to the dynamism of the post, a mix of new and old buildings, shabby and sharp.

The post trader's house reflected the highest aspirations of its owner (fig. 2-10). With a monopoly on commercial interests at the fort, the post trader tended to be wealthy. He set himself up as the social equal to the commanding officer, and used his genteel house to express his status. At Fort Laramie, the post trader's house had a symmetrical plan: two rooms separated by a central hall, with fireplaces located on the interior walls of these rooms. To this symmetrical plan rooted in classical style, decorative

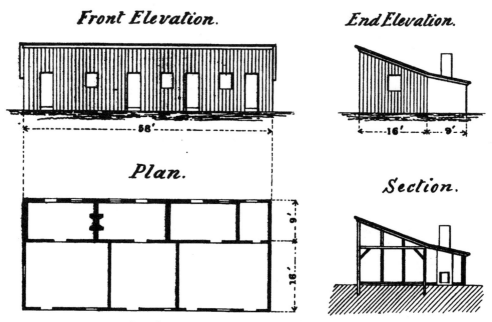

Fig. 2-7. Fort Laramie, laundresses' quarters No. 31, plan, elevations, and section, 1867. Laundresses usually received the least desirable quarters, such as these board-and-batten ones, built in 1858. Courtesy National Archives.

elements in the latest fashion, such as deep eaves ornamented with jig-sawn millwork, board-and-batten siding, and a steep gable roof, were applied, evoking the Gothic Revival mode. As Louis Simonin described it, the building was of "a style even stranger for this country, a kind of Swiss chalet." He compared it favorably to the post trader's store: "The refinement of this dwelling puts to shame the mean appearance of the low, gloomy canteen." Catharine Collins, wife of the post commander in 1863, attended a party in this "very pretty house": "The parlor is a beautiful though not large room with handsome curtains to the three windows, a beautiful brussels carpet, a few pictures and other nice furniture." The back wing contained the stairway to the second floor, a pantry, kitchen, and servant's room, while the second floor had three bedrooms. Constructed in 1860, the house was "built of yellow pine boards, lined inside with adobes, lathed and plastered," and sat on a stone foundation. Post traders Seth E. Ward and William G. Bullock valued the house at $5,000 when they offered to sell it to the army in 1866.[12] The sale did not go through, and a subsequent post trader replaced the house with a concrete one in the late 1870s.

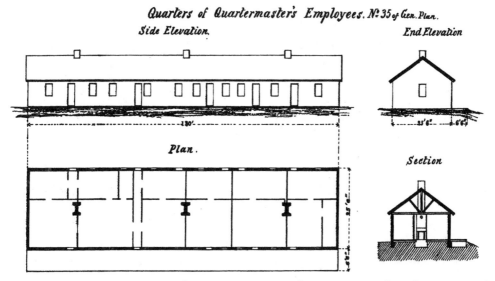

Fig. 2-8. Fort Laramie, quarters of quartermaster's employees No. 35, plan, elevations, and section, 1867. New in 1867, this six-unit adobe building housed civilian employees of the quartermaster's department. Courtesy National Archives.

As at Fort Laramie, officials at Fort Bridger did not start with a clean slate. When they abandoned Fort Bridger, the Mormons had left behind their stone fort, to which the army added lunettes during the winter of 1857–1858. The next summer, troops built a stone commissary storehouse in this compound. Most of the rest of the buildings were log. Taking advantage of extensive timber in the foothills of the nearby Uinta Mountains, the army built a complex of low, one-story buildings, using expeditious panel construction and dispensing with stone foundations.

On the east side of the parade ground stood a row of six similar officers' quarters (fig. 2-11). Despite the rough appearance of the log buildings, their center-hall plans and symmetrical arrangement of four rooms reveal a style-consciousness. Five of these were double officers' quarters, while the sixth was the commanding officer's quarters (fig. 2-12). It, too, had a center hall and four rooms, as well as a kitchen wing. (Kitchens were not added to the other officers' quarters until 1874.) The officers' quarters accommodated the complement of officers at the fort in 1858, which was eleven. The next year there were fourteen officers, but the following year, only five; the space allocated to officers would have shifted accordingly. Inside, the walls of the officers' quarters were lathed and plastered, while the barracks were only chinked. The plastering was a mixed blessing, however; one report stated that "the frost heaves and settles the buildings causing the plastering to crack and rendering them cold in the winter."[13]

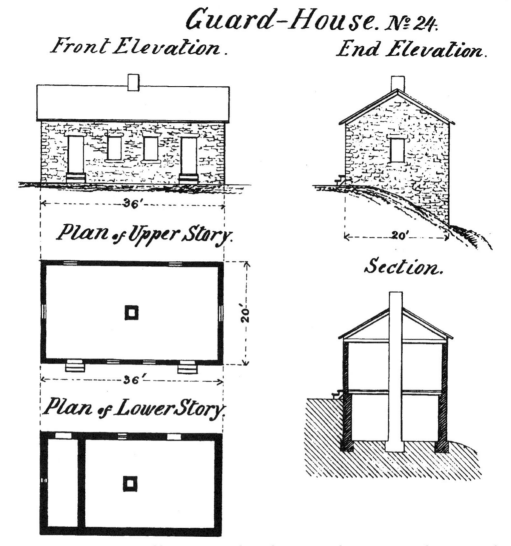

Fig. 2-9. Fort Laramie, guard house No. 24, plans, elevations, and section, 1867. The 1866 guard house was one of two stone buildings at the fort. Adobe was particularly unsuited to guard houses, as prisoners could dig their way to freedom. In this building, prisoners were kept in the lower level, while the top served for the guards of the day. Courtesy National Archives.

The quarters thus embody the contradictions in military architecture in the West. At first glance, they are low, crude, log structures, poorly built. But their center-passage plans and surviving fireplace surrounds, revealing a subtle hierarchy in room uses, demonstrate an architectural sophistication that links these log houses with eastern conventions (fig. 2-13).

Five barracks lined the north and south sides of the parade ground (fig. 2-14). Each one-story building measured 75 by 20 feet, demonstrating one of the advantages of panel construction: it could be extended indefinitely. The log barracks were constructed with no interior finish. In 1860 the

Fig. 2-10. Fort Laramie, post trader's house, ca. 1878. Post trader Gilbert H. Collins stands at the picket gate, with his wife and daughter on the porch of their Gothic Revival house, described by one visitor as "a kind of Swiss chalet." Courtesy the Bolln Collection, Fort Laramie National Historic Site.

commanding officer proposed to line them with canvas. Misunderstanding the request, the deputy quartermaster general assumed that Maj. Edward R. S. Canby wished to lath-and-plaster the barracks; this was disapproved. He did, however, permit the ceilings to be finished with boards, but this, as well as flooring for the barracks, had to wait for the government sawmill, due to arrive that fall. In 1866, however, Fort Bridger was still without a sawmill, although the post trader had one. The post also obtained lumber, shingles, and lath from Salt Lake City, 120 miles away.[14]

The barracks remained porous to wind and cold. In 1874 Assistant Surgeon Charles Smart conducted extensive experiments at Fort Bridger, measuring the air in various barracks at different times through the night. He also calculated the disease rate in different barracks occupied by troops of the 4th Infantry. Company B, which had the highest rate of disease, occupied barracks that were in good condition, having been repaired just the year before. Company C experienced half the disease rate of Company B; C's barracks had been "plastered during the past autumn, but much of it has already fallen. About one-sixteenth of the ceiling . . . is denuded,

Fig. 2-11. Fort Bridger, officers' quarters, 1866. Double officers' quarters built in 1858 employed the expeditious panel-log construction. Courtesy Wyoming State Archives, Department of State Parks and Cultural Resources.

showing the laths in circular patches, and through these and crevices in the roof above daylight can be seen. . . . A broken window-pane was found in each room." Supporting contemporary theories about ventilation, the deteriorated but better ventilated barracks produced fewer diseases. Post surgeons managed to see the bright side of barracks with gaps and cracks: they facilitated ventilation. The temperature fell to −20°F. that winter; that year there were only two months when the temperature did not drop below freezing at Fort Bridger.[15]

Behind the barracks, mess rooms and kitchens occupied separate buildings (fig. 2-15). Laundresses' quarters, 20 by 100 feet, were located on the east end of the south row of barracks, facing onto the parade ground. This was a most unusual arrangement, for laundresses were usually banished to unfavorable spots off the parade ground. The laundresses' quarters resembled the barracks, being constructed of the same paneled logs in roughly the same plan.

The row of officers' quarters and the rows of barracks represented an orderliness and uniformity valued by the army but rarely attained in the architecture. Constructing eleven log residences, a hospital, eleven storehouses, and a blacksmith shop all in the summer of 1858, however, stretched the fort's 275 officers and men to the limit. The men had to use expeditious construction methods, which resulted in the need for repairs. In 1873, just fifteen years after original construction, all the barracks and

HUTS AND ADOBE BUILDINGS SADLY IN NEED OF REPAIR ■ 57

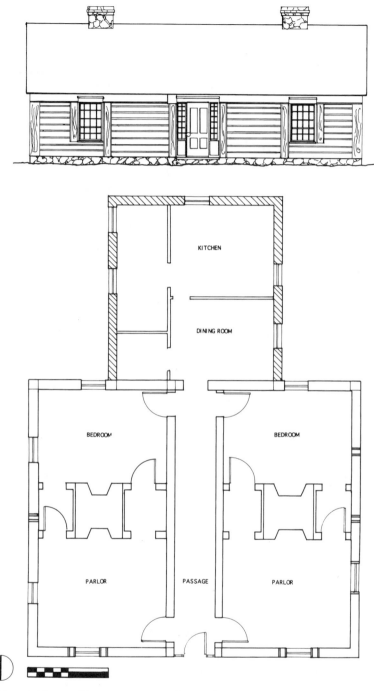

Fig. 2-12. Fort Bridger, officers' quarters, plan and elevation. A wide center hall divided the log officers' quarters into two units or permitted it to be used as one. The frame kitchen wing, implying shared dining facilities, was not added until 1874. Drawing by Heather Randall, 1996, Western Regional Architecture Program, Graduate School of Architecture, University of Utah.

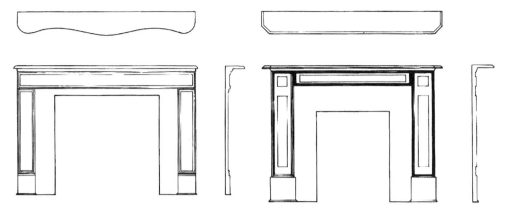

Fig. 2-13. Fort Bridger, officers' quarters, detail of fireplace surrounds. The more elaborate decoration of the fireplace in the front room, right, compared to the room behind it, indicated that the front room was the more public space. Drawing by Liza Hart, 1996, Western Regional Architecture Program, Graduate School of Architecture, University of Utah.

quarters received extensive renovations, including new flooring, plastering, and shingling. The roofs of the barracks, which had yielded a ceiling height of 7½ feet, were raised to provide a 9-foot height, improving the air circulation deemed so important by the surgeon general's office. Porches were added to the barracks in the 1860s, providing shelter to the doors and windows as well as an informal gathering place.[16]

Personnel

One of the constraints placed upon the production of decent architecture at western posts was the personnel available. Responsibility for the construction of buildings at a post rested with the acting assistant quartermaster, also called a post quartermaster, who was a line officer assigned to the task. This lieutenant often had little training as a quartermaster and was burdened with many duties. The responsibility of the quartermaster department that devolved onto this acting assistant quartermaster included provision not only of housing, but also of transportation and clothing. His daily correspondence ranged widely: arranging for shipping, disbursing extra-duty pay, hiring civilians, issuing stationery and postage stamps, and offering rewards for the capture of deserters. By the 1880s it also included arrangements for coal, lamp oil, telephone service, and bunks. The post quartermaster's work involved translating other officers' wishes into requests directed to the appropriate authority and handling the disbursal of money once permission to acquire something had been granted. Even Quartermaster Gen. Montgomery C. Meigs called the post quartermaster's duties "responsible and onerous," and pointed out that the assignment was

Fig. 2-14. Fort Bridger, barracks, undated. Panel construction also worked well for larger buildings, such as these barracks. Five such barracks lined the parade ground. Courtesy Wyoming State Archives, Department of State Parks and Cultural Resources.

"avoided rather than sought." Given this kind of reputation for the job, it was little wonder that the post quartermasters had high turnover. Meigs determined that the average number of post quartermasters in 1871 was 150, although the number who served in that capacity was 443. Given that so many were new to the job, the post quartermasters relied on quartermaster sergeants, clerks who stayed in that position for longer periods of time. Beginning in 1884, a quartermaster sergeant was assigned to each post.[17]

Although it formed only a part of the post quartermaster's duties, construction of new buildings was a complicated process. Before the secretary

---

"Sorrows of an A.A.Q.M. [Acting Assistant Quartermaster]"

This article was signed with the classical pseudonym "Caius" in the *Army and Navy Journal* in 1869. The writer's actual name and identity are unknown.

Your easy careless life is at an end. You are elevated to a distinction of which you never dreamed unless like Hamlet you have had bad

> dreams. Everybody wants you. Children cry for you. The general sends for you to inquire whether you made that requisition; prepared that estimate; sent that team; completed those plans; wrote that letter. Or else he desires information as to why that ambulance didn't report; why those stores haven't arrived; why you haven't got better army clothing; why that gate isn't hung; why some delectable extra-duty man wasn't at his proper place, etc. . . .
>
> Upon some rainy day, you may have allowed your carpenters, in default of their ability to work outside, to make something for Mrs. X. Upon a fair day when every hammer is busy, every saw flying on post repairs, Mrs. Z. sends to you for a carpenter to make for her immediately a work-table. You send back a polite message explaining the present impossibility of the thing; Mrs. Z. denounces you as partial.
>
> You have a little walnut lumber which you obtained to construct coffins for any unfortunate military heroes of the garrison who may depart this life. Out of this you are desired to furnish material for everything, from a picture-frame to a set of furniture. With a small force of carpenters you are expected to successfully conduct all of the work of repairing public buildings and also to make for the various benedicts of the post, hencoops, hat-racks, cradles, dining tables, etc., and also for some covetous bachelor sub[altern], an elaborate chest or two, which the subs desire. . . .
>
> The versatile learning required in the [quartermaster's] department, would make a man successful as a clothing merchant, or a veterinary surgeon; as an architect, or a farrier; as a livery-stable proprietor, or a surveyor; as a local editor, or a painter and glazier.
>
> Source: *Army and Navy Journal* 7 (11 December 1869): 258.

of war granted permission to build and allocated money to proceed, the post quartermaster had to produce drawings and estimate costs of the new building. To estimate costs accurately, he had to know who was going to do the work and where the materials would come from, and at what cost. Once he had this information, together with a design of the structure, he sent it forth, beginning with his post commander, for various approvals, which came in the form of endorsements written onto his request. For example, when Fort Laramie's post quartermaster, 1st Lt. George A. Drew, requested approval for construction and repairs on April 25, 1879, the request went first to the Cheyenne Depot, then received two endorsements at the Headquarters, Department of the Platte, in Omaha. The request then

Fig. 2-15. Fort Bridger, barracks, ca. 1889. The view northwest across the fort exposes the messier back sides of the barracks, with privies, mess halls, and kitchens, all constructed of logs. On the far side of the parade ground, the large building near the center is the frame barracks built in 1883, and to its left, the stone barracks built four years later. Courtesy Wyoming State Archives, Department of State Parks and Cultural Resources.

moved to Chicago, where it received two endorsements from Headquarters, Military Division of the Missouri. In Washington, D.C., it gained endorsements from the adjutant general, the quartermaster general, and the chief clerk of the War Department, who finally granted his approval on August 8. These eight endorsements were relatively simple; one request for an addition to the hospital at Fort D. A. Russell required forty endorsements, as the request was returned down the line to the post surgeon with questions, then passed back up again.[18]

The designs of the buildings originated with officers at the post or department level. The departments included a chief quartermaster and assistant quartermasters, all staff positions. If a significant building program was envisioned, the chief quartermaster might detail an assistant quartermaster—someone from his staff—to a particular post. At the post were usually only officers from the line; the post commander or the acting assistant quartermaster would be the most likely candidates to design buildings. Both were often inadequate to the task. Col. Henry E. Maynadier, commanding the District of the Platte, headquartered at Fort Laramie just after the Civil War, asked for a "quartermaster of experience," as "the present quartermaster of the post of Fort Laramie has not sufficient experience or knowledge of business."[19] A few months later, he received George B. Dandy, appointed chief quartermaster of the District of the Platte. When the District of the Platte disappeared as the result of a reorganization soon thereafter, Dandy stayed on at Fort Laramie as assistant quartermaster and oversaw building operations for another year.

One post commander with a strong interest in architecture was Col. Henry B. Carrington, who claimed credit for the log construction system his men used at Fort Phil Kearny in 1866. After his retirement, Carrington taught architecture, as well as military science, at Wabash College, and designed several buildings in Crawfordsville, Indiana. Not every officer was such an accomplished designer. Inspector General Randolph B. Marcy criticized the architectural plans of the commanding officer at Fort Reynolds, Colorado. Marcy called the plans "unsuitable" and said that Capt. Simon Snyder "seemed to me to evince no qualifications for directing such a work." As another observer noted, "At almost every post we have buildings which are monuments, more or less colossal, of the general stupidity of some post commander or post quartermaster, when their energy or whims prompt them to try their hands at an unlearned profession."[20] As this commentator pointed out, few designers of either the staff or line had any architectural training.

Although the numbers varied, through most of the nineteenth century about half of the army's officers had been educated at West Point. In the mid-1850s, 73 percent of the officers were West Point graduates, 25 percent were drawn from civilian life, and only 2 percent from the ranks. By 1883, 43 percent of the officers were West Point–educated, 49 percent came from civilian life, and 8 percent from the ranks.[21]

The academic training provided at West Point emphasized military and civil engineering. The well-respected professor Dennis Hart Mahan taught this course from 1830 until 1871, and one of the many texts that he published was *Notes on Architecture and Stone Cutting*. The thrust of the engineering course, however, was toward permanent, usually stone, fortifications. The number of lessons in the civil engineering course devoted to architecture was two in the 1840s, six in 1860, and none in 1865. In his curriculum for the 1849–1850 school year, Mahan described the architecture lessons: "Architecture. General nomenclature and divisions of. General observations on the various styles of architecture and their orders. Description and uses of the elements of edifices as walls, columns, roofs, floors, windows, doors, flues, etc. General principles of proportions of the various parts of edifices. Analysis of the general and particular conditions of suitableness in all structures of civil edifices." Little attention was given to temporary forts in the West, just as little discussion of military strategy at West Point was devoted to the unconventional warfare that took place in the West. Instead, cadets learned about the construction of arches, bridges, public works, and fortifications.[22] When young lieutenants found themselves in charge of new construction at western forts, they had little experience with architectural design or construction supervision. Even the staff officers assigned to the department had little training, although they benefited from more experience.

The troops themselves usually constructed new buildings, as authority to contract out a construction job was granted only for permanent posts. In the West, construction tasks overwhelmed all others, to the detriment of morale. In 1870 General Sherman objected to using soldiers as laborers, pointing to high desertion rates. As one observer noted, "This 'labor of the troops' was a great thing. It made the poor wretch who enlisted under the vague notion that his admiring country needed his services to quell hostile Indians, suddenly find himself a brevet architect, carrying a hod and doing odd jobs of plastering and kalsomining [calcimining]." A group of soldiers complained to Congress in 1878 that their duties included "all the operations of building quarters, stables, storehouses, bridges, roads, and telegraph lines; involving logging, lumbering, quarrying, adobe and brick making, lime-burning, mason-work, plastering, carpentering, painting, etc." For this work, soldiers expected to receive extra-duty pay, ranging from 25 cents to $1.00 per day, in addition to their regular wages of about $13 per month. Soldiers did not necessarily have any experience in the various construction jobs they had to do. Of fifty-four recruits at Fort D. A.

Russell in 1883, for example, one had been a bricklayer and one a painter.[23]

Civilians with construction expertise often supervised troop labor. The acting inspector general of the Department of the Platte observed in 1866: "I have to recommend that a few hired artisans be allowed at each post; their pay will be saved in material; and their work will be doubly efficient." Sometimes, however, the employment of skilled civilians failed to save the army time or money. For the construction of the new post of Fort D. A. Russell in 1867, the army hired six brick masons at $100 per month. An officer described the construction work: "The building was not commenced until late in the season, and instead of giving it out by contract, is being built by the quartermaster's department"—in other words, by workmen hired by the quartermaster's department, paid by wage, not by the job. "Every man employed is interested in having the job last all Winter, and I think from present appearances they will succeed in their intention." The troops, too, were working: "My company have been hard at work, assisting in building quarters for themselves, in fact, every man at the post has been so employed."[24]

In 1866 at Fort Laramie, Assistant Quartermaster Dandy had close to a hundred soldiers at his disposal. Sixteen served as masons, carpenters, saddlers, blacksmiths, messengers, whitewashers, plasterers, or painters; sixteen were teamsters; twelve worked at the sawmill; eleven burned lime; ten quarried stone; and twenty-three cut wood. Dandy requested that fifty additional soldiers be detailed as laborers, but the post adjutant replied that no more men could be spared. Dandy appealed to the chief quartermaster at the department, who the following spring hired in Omaha several "mechanics," including twelve carpenters and one painter, to work at Fort Laramie for three months, receiving in addition to their pay one ration per day and transportation back to Omaha at the end.[25]

Post commanders and acting assistant quartermasters easily recognized the advantages of hiring laborers. The civilians provided needed skills and allowed soldiers to do soldiering, instead of building tasks. But limited budgets made civilian labor something of a luxury. In 1866, when Meigs received estimates for construction that he found unreasonable, he objected to Commanding General Grant and recommended "enforcing the discharge of civilians, and the employment of troops in all labors and work relating to service of supply and shelter."[26]

Space

Like logistical and personnel issues, the question of space imposed constraints on army architecture. The amount of space to which a soldier was entitled, according to rank, as spelled out in official army regulations, was the only clear guidance that a post quartermaster received when designing a building. It was also the one aspect of the architecture that caused the most complaints, especially on the part of officers and their wives. Although these regulations stayed on the books until 1908, they were often disregarded.[27]

Regulations published in 1855 set the space available to enlisted men and noncommissioned officers at 225 square feet for every six men when they were housed north of 38°N latitude and 256 square feet south of that latitude. In practice, the men received whatever space was available. Each company was assigned its own barracks, so if the size of a company diminished, more space became available for each person. Changes in sleeping practices also affected space in barracks. Until after the Civil War, the army practiced double-bunking—that is, sleeping two to a bunk. Not only were there two men to each wooden bunk, but the bunks themselves were two-tiered. During the Civil War bunks were often stacked in three tiers, and this remained an easy way to increase barracks capacity, as Assistant Quartermaster Dandy did at Fort Laramie in 1866. The army introduced single, iron bunks in 1858, but they were not distributed widely until after the Civil War.[28]

The regulation of square footage for enlisted men was predicated on the assumption that they would be sharing quarters, but not necessarily in one open room. The space allocation of square footage for every six men implied that a smaller group than a full company would occupy one room. The 1849 barracks at Fort Laramie slept a company of seventy-four men in four 16-by-30-foot rooms on the second floor. The first floor was devoted to the kitchen, mess hall, and rooms for noncommissioned officers. Similarly, barracks at Fort Bridger built in 1858 had multiple sleeping rooms. An interest in the beneficial effects of ventilation drove the shift toward large, open dormitories, preferably no more than 24 feet wide, with windows on both long sides.[29] With this arrangement as a constant, barracks did not differ widely in appearance. The dormitory was generally a long, narrow room, stretched across the front of the building. The mess hall would be located either in a wing in the rear, forming an L-shaped or T-shaped plan, or in a separate building in the rear. In two-story barracks, the second floor was devoted to the dormitory, while the first floor held the mess hall and noncommissioned officers' rooms. Barracks had gable or hipped roofs and were usually the widest buildings fronting on the parade ground, with porches stretching across the front.

Officers' quarters varied in appearance, exhibiting a deliberate hierarchy. Higher ranking officers received more rooms, so the size of the units announced the rank of their occupants. In addition to the ubiquitous kitchen, army regulations fixed the allocation at one room for lieutenants, two rooms for captains, three rooms for majors and lieutenant colonels, four rooms for colonels and brigadier generals, and five rooms for a major general. This was an adequate allotment for unmarried officers, who often shared quarters or at least messed together. Officers with wives, children, and live-in servants, though, found their quarters insufficient, and they did not hesitate to complain. One described "this one room and a kitchen existence" from an unmarried lieutenant's point of view: "To live in one's bedroom, to receive one's friends there; to smoke socially in the same apartment where one is to sleep, to breakfast and dine in one's kitchen. How delightful! How it would tempt us to sit over our cigars and our wine while Bridget is occupied washing the platter, or in banishment wait outside." He then looked at it from a prospective husband's viewpoint: "Fortunately spoony people never let such sublunary affairs as kitchens enter their thoughts. If they did, and the young woman was at all matter of fact, there might be fewer unions and more broken-hearted subalterns."[30]

In practice, officers did not always receive their regulated allotment of rooms; it depended on what was available at a post. The assignment of quarters was based on rank, so that when an officer was sent to a new post, he could select the best quarters for which he was qualified. If they were occupied by someone of lower rank, then that officer would have to move out, but could bump someone of lower rank than himself, and so on. Thus, in 1891 at Fort D. A. Russell, Maj. Valery Havard selected quarters No. 30, displacing Capt. Henry S. Howe, who selected quarters No. 28, displacing the post quartermaster, 1st Lt. Edward Chynoweth, who selected quarters No. 17, displacing 2nd Lt. Lucius L. Durfee, who selected quarters No. 27.[31]

Laundresses also qualified for quarters on the post, at the same rate as enlisted men—in other words, 225 square feet for every six laundresses. As laundresses did not live together in barracks, but resided in individual quarters with their children and sometimes their soldier-husbands, this was an absurd allocation, one that clearly was not meant to be obeyed. In practice, the laundresses' low status at the fort was reflected in the poor quality of their housing, usually the worst at the post, far off the parade ground.[32]

Building Materials

The room allocations provided a minimum, a level below which post quartermasters were certain they could not go, so when they looked for ways to cut costs, they had to look elsewhere. One of the few ways that an acting assistant quartermaster could stretch a limited budget was through the choice of construction materials. As transportation of materials was difficult and costly, the army tried to rely on available materials to a great extent, proving its flexibility and willingness to adapt. Long accustomed to constructing wood-frame buildings, the army was suddenly confronted with a situation in which that was impossible, because in most of the Great Plains, there was not sufficient timber to make wood-frame construction a realistic possibility. The army accordingly turned to solutions that other newcomers to the region had already discovered, such as adobe.

In the first two decades of Fort Laramie's existence, supplies came, like the emigrants, on the overland trail from St. Louis. Fort Laramie was 1,000 miles from St. Louis and 600 miles from Omaha; 20 miles a day was a good pace for a wagon train. Once the transcontinental railroad was established, freight was shipped by rail to Cheyenne, then overland 89 miles north. Even if most construction materials were locally obtained, items such as window glass, hardware, and stoves still had to make this arduous journey. In 1875 the post quartermaster at Fort Bridger estimated the materials he needed for new construction, but was unable to price several of the items, such as spikes, nails, brick, tin, plastering hair, white lead, linseed oil, turpentine, wallpaper and border paper, and glass, because they were not available locally.[33] The chief quartermaster in the Department of the Platte purchased these supplies in Omaha and shipped them on the railroad, which by then came within 10 miles of Fort Bridger.

Ideally, the buildings at a post would be constructed of masonry, projecting the proper military image of permanence and impermeability. But for reasons of cost, stone and brick were generally out of the question for temporary western forts that might be abandoned after five or ten years. The army commonly turned to wood frame, especially in timbered regions. All of the eastern military men were familiar with wood frame, covered with either board-and-batten or clapboards, and a post with such construction appeared familiar and natural to them.

Even for forts located in timberless regions, wood was still essential to construction, if only for floors and roof framing. Commanding officers tried to obtain a sawmill soon after selecting the site of a new post; Fort Laramie had a sawmill by November 1, 1849, just months after the government acquired the fort. The sawmill was a constant source of problems, several times suffering damage by fire (fig. 2-16). As described in 1866, "the saw mill is working slowly, being old, shaky, and needing constant repairs." It was located near timber, some 55 miles from Fort Laramie, over a bad road. "A party of hostile Indians attacked the mill a few days ago, ran off six mules, and drove the party at the mill into their stockade." Sawmills could be powered by animals or steam, although the latter required an

Fig. 2-16. Fort Laramie, sawmill at Laramie Peak, drawing by Caspar Collins, ca. 1863. Located in the nearest wooded region, some 55 miles from the fort, the steam sawmill was staffed by a contingent of troops. Courtesy American Heritage Center, University of Wyoming.

engineer or someone with some skill to operate it. One observer criticized the civilian sawmill operator at Fort Phil Kearny: "If Colonel Carrington fails with his mills, it will be from the inferiority of his engineer, whom he picked up along the way." In 1863 the sergeant detailed as an engineer to operate the sawmill at Laramie Peak met an unfortunate if ironic end when, as he was visiting the fort, an adobe wall collapsed and killed him.[34]

Sawmills processed both wood for fuel and lumber for construction; highly finished millwork usually came from elsewhere. Maj. William Hoffman, commanding Fort Laramie in 1856, complained that local lumber warped excessively, so that doors, window sashes, and shutters had to be shipped from St. Louis. In 1867, dressed flooring, lath, and finished lumber were shipped from Denver to Fort Laramie, a journey of 210 miles that took, in one case, forty-three days. By 1873 the post quartermaster, 1st Lt. Charles H. Warrens, assured authorities that "all the material required for the repairs or erection of buildings can be manufactured or obtained at or near the post, with the exception of flooring." Local lumber was not suitable for flooring, nor did Fort Laramie have tongue-and-groove machinery.[35]

Wood also served as a roofing material. Major Hoffman requested a shingle mill—even one that could be attached to the sawmill—because the "ordinary method" of making shingles by "splitting and shaving" would not work, presumably due to the quality of the timber. Hoffman apparently received his shingle mill, because a shingle mill was reported damaged by fire in 1859. In 1866 shingles en route from Denver were delayed on the road, taking about two months to arrive, leaving half-built adobe buildings exposed to the rain. Without a shingle mill, soldiers made shingles by hand, as described by Margaret Carrington at Fort Phil Kearny in 1866: "Shingles were rived from bolts sawed by the men, and many a 'shingle bee' was held, at night, to expedite work and convince the skeptical that shingles, or anything else, could be made or *done*, when it *had* to be, and that civilization was still westward bound."[36]

Whether or not posts were equipped with sawmills, wood was still an easily worked construction material. Fort Bridger was located in a timbered region near the Uinta Mountains. The buildings constructed in the first year of the army's possession in 1858 were log, except for the commissary storehouse, which was stone; the post did not receive a sawmill until 1860. The army, like others establishing settlements in remote regions, favored log construction; at its most basic level, this required no sawmill, but only an axe. One observer of military architecture noted that the strength of a log house depended on its corners, citing dovetail, saddle-notching, and "halving and pin"—all traditional cornering techniques. But forming corners by traditional techniques was time-consuming, as each corner notch was created for individual fit. For construction in quantity, the army was forced to adopt the more expeditious system of panel construction, which involved horizontal logs laid in between hewn vertical posts. The most likely source of paneled construction was French Canada, where the technique was called *pièce sur pièce* and was used commonly. From Canada, it spread to the northern United States and throughout Wyoming, where it was used at scattered ranches as well as in South Pass City.[37] All of Fort Bridger's log buildings employed the slot-and-cleat method of panel construction.

Panel Construction

Panel construction involved setting horizontal logs into vertical posts in approximately 8-foot sections, or panels. Lt. Col. Thomas M. Anderson identified three ways to join the panels to the posts: mortise and tenon, slot and cleat, or the toe-nail. He dismissed as unsuitable only the toe-nail, in which the horizontal logs were nailed into the corner posts, such as were used at Fort McKinney, Wyoming. A fourth method, not mentioned by Anderson but employed at Fort Robinson, Nebraska, was the hog-trough, in which horizontal logs were nailed into vertical boards at the corners. The ready availability

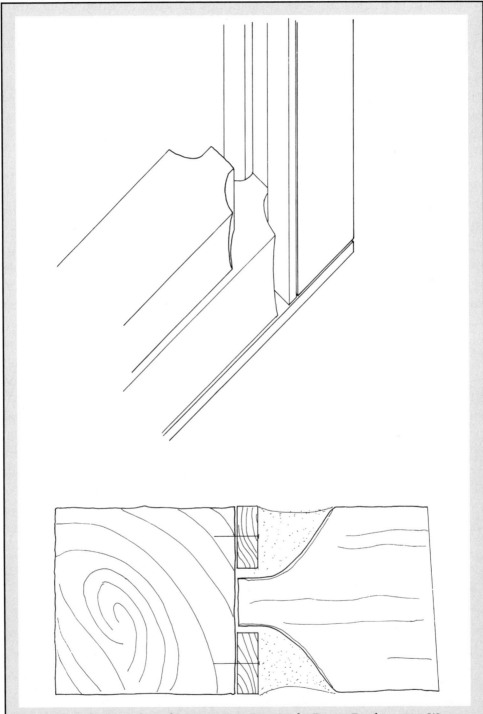

Fort Bridger, diagram of panel construction. Drawing by Barrett Powley, 1996, Western Regional Architecture Program, Graduate School of Architecture, University of Utah.

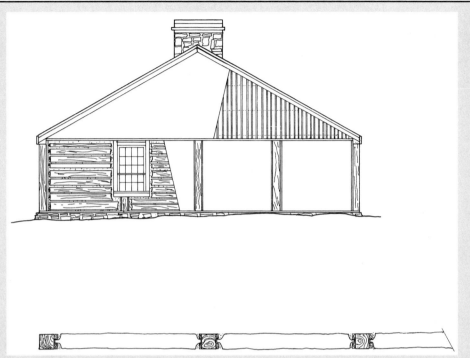

Fort Bridger, log officers' quarters and plan of panel construction. Drawing by Barrett Powley, 1996, Western Regional Architecture Program, Graduate School of Architecture, University of Utah.

of nails is reflected in the toe-nail and hog-trough methods, both less stable and durable than the other two techniques.

At Fort Bridger, troops used the slot and cleat, in which 7- to 8-foot-long, square-hewn logs were narrowed at the ends, laid horizontally, butted against vertical posts, and held in place by smaller vertical pieces that formed a slot. In 1859 Major Canby described log officers' quarters constructed there: "Buildings are of pine logs empaneled into a frame and supported by cleats." Workers at Fort Bridger constructed six officers' quarters and five barracks using this system, which offered numerous advantages. Little skill was required: all of the pieces could be standardized rather than custom-fitted, as required by dovetail or saddle-notching. Construction would be simple, fast, and cheap. Furthermore, in this essentially modular system the walls could be extended indefinitely, independent of the length of the timber.

Mortise-and-tenon panel construction required that the horizontal logs be mortised into the vertical posts. Col. Henry B. Carrington illustrated this system in his description of Fort Phil Kearny in 1866. He noted that no "citizen-mechanics" were allowed, and the only tools required were a broad-ax, 2½-inch auger, and chisel. Soldiers hewed the logs flat on two sides, "to dispense with spikes." They set the vertical posts in the ground, and chinked

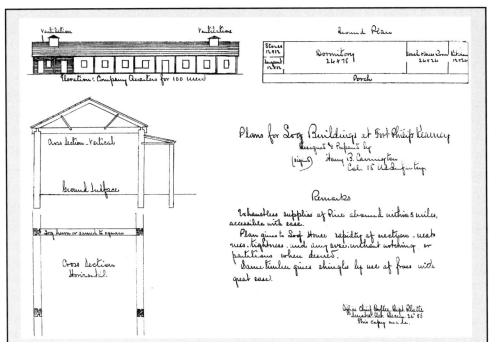

Fort Phil Kearny, log buildings, plan, elevation, section, 1866. Courtesy National Archives.

the walls with clay. After his steam sawmill arrived, Carrington ordered the logs squared on four sides, creating shingles out of the discarded lumber. This construction system provided "rapidity of erection, neatness, tightness, and any size, without notching or partitions, when desired." The mortise-and-tenon system was used again at Fort Sanders, near Laramie, in 1867, where six "citizen carpenters" aided construction. The fort was described as follows: "The buildings are all of regular panel work, 13 to 14 feet, let into mortises in posts, 8 [inches square] by 8 feet [tall], the posts being thoroughly pinned, top and bottom, through plate, girder and sill."

Sources: Anderson, "Army Posts, Barracks and Quarters," 440. O'Dell, "Building Technology in the Department of the Platte, 1866–1890," 42. Canby quote on drawing in RG 77, Miscellaneous Forts File—Fort Bridger, #20, Cartographic Division, National Archives, College Park, Maryland. On mortise-and-tenon construction, see text accompanying drawing in RG 393, Part I, Entry 3898, Box 15, NAB. *Army and Navy Journal* 4 (5 January 1867): 313.

The timberless nature of much of the West increased costs. As Quartermaster General Meigs explained, "The shelter of the troops in the treeless regions now occupied is more costly than it was when the frontier posts were in thickly wooded districts. There are places in Texas and Arizona, and on the plains, where the timber necessary for roofing and flooring is hauled by wagon trains hundreds of miles." When the U.S. Army acquired the trading post at Fort Laramie in 1849, the troops obtained timber for the sawmill 12 miles away, but within two years they had to go 25 miles for mill lumber. By 1856 they went 35 miles; by 1863, 45 miles; and by 1866, 55 miles, deeper into the Laramie Mountains.[38] A substitute for wood was sorely needed.

When the U.S. Army acquired Fort Laramie, the existing trading post proposed its own solution to the timber problem: adobe. Introduced by Spanish colonists in the early seventeenth century, this construction system spread northward. Lt. Daniel Woodbury of the Engineer Department had supervised extensive adobe construction at Fort Kearny in 1848; he gained his experience from a trader named Andrew Sublette, who had built Fort Vasquez, a trading post in Colorado, out of adobe in 1835.[39] Woodbury supervised the occupation and repair of the adobe trading post at Fort Laramie, as well as the construction of new buildings, some of which were adobe. By February 1851 troops had constructed four new adobe buildings.

The appeal of adobe to the army was plain. It was an indigenous material that required little skill to fabricate. Simply, men mixed earth with water, and sometimes straw, to form a material that they molded into bricks about 9 by 18 inches and 6 inches thick, which they left to dry in the sun. Unlike brick, adobe did not require a fuel source to fire it. The adobe bricks were laid up like masonry, then covered with a mud plaster. The post surgeon at Fort Laramie noted in 1869 that "a party of six men are employed on sunny days at making adobes for building purposes." When several companies constructed a new site for Fort Thornburgh in northeastern Utah in the summer of 1882, one unhappy soldier reported that "a part of the command, with a saw mill, was sent to the timber, 20 miles distant, and have been and are still getting out lumber; the balance have been kept busy making adobes, of which about 180,000 were manufactured during the summer. As yet [September 18] quarters have not been commenced for the men."[40]

Major Hoffman, commanding Fort Laramie in 1856, found adobe particularly appropriate for an unskilled workforce: "From the experiments I have made here, I am satisfied that the adobe is the best, and much the cheapest material, that can be used for buildings in this country. Last summer I commenced building in August, with only three rough masons and not a man who had ever made an adobe and by November seven adobe buildings were up and occupied. Any handy man can be taught to make, and to lay adobes in a wall, in a very little while." During a construction boom in 1866, however, the assistant quartermaster at Fort Laramie contracted for 250,000 adobe bricks; apparently troops could not supply the demand.[41]

Adobe served well for building purposes, providing sturdy and durable walls. As Lt. Col. Thomas M. Anderson described it, "adobe houses with thick walls, are warm in winter and cool in summer."[42] The insulating property of adobe bricks was such that they were used as infill for wood-frame buildings, such as the two-story officers' quarters built at Fort Laramie in 1849 and the barracks at Fort D. A. Russell built in 1867.

Despite its initial effectiveness, adobe needed frequent maintenance, and this caused it to be often equated with shabbiness. As Secretary of War William Belknap put it in 1874, "The Army is in many localities badly sheltered, living in huts and adobe buildings sadly in need of repair." Adobe requires annual patching and occasional replastering; soldiers may have been unaware of this need or unwilling to take on this work. As Col. Henry Douglass described Fort Union, New Mexico, in 1886, "the adobe buildings at this post were originally plastered on the exterior to protect them from the washings of the violent storms which prevail here during the summer. This plaster has fallen off, leaving the walls exposed to the weather." In 1873 Quartermaster General Meigs denied a request to hire outside labor to fix the buildings: "I do not think it impossible for a company of American troops to take care of their own quarters. The people of a Mexican village, less educated—not more apt—build their village without recourse to the outside world, and even provide all the material, to shelter themselves."[43] Soldiers, however, failed to maintain their adobe buildings, so that adobe remained associated with poor quality.

Effect of Constraints on Architecture

The use of local materials, soldier labor, and officer designers was a stopgap solution to the problem of inadequate funding. Quartermaster General Meigs recognized this problem. Almost annually, he pleaded for a greater appropriation for the "barracks and quarters" account. In 1870 he claimed that his office had "constantly endeavored to improve the condition of the quarters and the accommodations of the Army," but blamed his failure on the inexperience, inability, and frequent moves of the acting assistant quartermasters. The next year, he seemed resigned to low quality, pointing out that these were temporary posts built by the labor of the troops and that they would necessarily be minimal accommodations: "The difficulty is, that our Barracks must be, generally, built hastily, in the wilderness, where material and labor are both scanty and costly, and where men live exposed to the inclemency of the weather, till the barracks—erected in great part by their own labor—provide them with shelter. Shelter is all that they are willing to give the necessary labor to provide." He wrote somewhat longingly of quarters "in Havana, and in European cities, where some of the Barracks are Palaces in extent and in beauty," but noted that Congress was not willing to fund these except in the case of permanent fortifications, which were built by the Engineer Department, not Meigs's Quartermaster's Department.[44]

Congress's reluctance to improve posts in the West took its toll on the quality of the housing. By 1874 the secretary of war was again pleading for funds: "As long as the Army is in many localities badly sheltered, living in huts and adobe buildings sadly in need of repair, the roofs leaking, the walls open to the inclement weather, I must repeat what I have so often insisted upon, that the appropriation for 'barracks and quarters' . . . [is] entirely inadequate to the necessities of the service and the health and comfort of our troops." As the post surgeon at Fort Fetterman described its buildings: "Here, as almost everywhere, the results of green lumber and army mechanics are visible in wide cracks in roof, floor, window and door frames, admitting rain, draught, and (worst of all) dust."[45] The poorly built and poorly maintained quarters had a detrimental effect on the health, comfort, and morale of the troops.

As grand as the army's plans for a fort might be, they still faced serious constraints, so that the resulting collection of buildings at a fort embodied contradictions. Sometimes the contradictions fell within a building type, such as officers' quarters at Fort Laramie that ranged from single-story linear structures to two-story buildings with passages. Sometimes the contradictions occurred within a single building, as in the log officers' quarters at Fort Bridger, which had a genteel Georgian plan in a rough-hewn log building. Although glimpses of the army's high architectural ideals can be found in the pair of two-story buildings erected at Fort Laramie in 1849–1850, the mitigating factors of budget, supply, and labor pool would make such buildings increasingly difficult to realize.

PART II | **VILLAGE**

Forts Laramie, Bridger, and D. A. Russell, 1867–1890

3

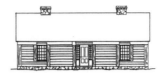

A BEAUTIFUL VILLAGE

Purpose and Plan

In 1870 Lt. Col. Albert G. Brackett described Fort Bridger as "a beautiful village." When Elizabeth Burt returned to Fort D. A. Russell in 1868, she reported that "frame quarters for a regiment had sprung up as it were, into a small village." Frenchman Louis Simonin said of Fort Laramie in 1867 that "the fort resembled more a Spanish-American village than a military post of the United States." The village analogy served even for generic descriptions of western posts; in a reminiscence of his soldiering days, George Forsyth described a fort from the point of view of a civilian visitor: "Instead of bastioned walls, deep ditches, and grassy ramparts, from which frown deep-throated cannon, he sees before him, as he leisurely approaches, what at first sight appears to be small village set well out in the plain or possibly at its edge near an outlying mountain generally embowered in shade trees, with a tall flagstaff in its centre, from which floats the flag of his country."[1] "Village" is the way that forts struck visitors at the end of the 1860s, and "village" described the character of western forts for the next twenty-five years. A village may not have been what army officials intended to build in the West, but nevertheless it was an image with which they were comfortable.

When nineteenth-century observers used the term *village,* they were probably thinking of the New England, Pennsylvania, or Ohio communities that were the birthplaces of most officers (fig. 3-1). In identifying symbolic American landscapes, geographer D. W. Meinig named the New England village as one whose meaning—its association with community values—as

Fig. 3-1. Concord, Massachusetts, drawing by J. W. Barber, ca. 1835. By the mid-nineteenth century, the New England village evoked a scene of large white buildings clustered around an open space lined with trees. John Warner Barber popularized this image through engravings such as this. Courtesy American Antiquarian Society.

well as appearance are assumed to be known. Given the phrase "New England village," Americans instantly picture a village green surrounded by white frame houses, marked by a white, steepled church, and richly planted with great trees. Historian John Reps traced the New England village to seventeenth-century settlement patterns in which farmers lived in compact settlements, while their farms extended beyond. Enumerating the features of a village, Reps mentioned the presence a village green, a careful siting of buildings, a sharp break between village and countryside, and a settlement planned for a limited population. Geographer Joseph Wood argued that the stereotypical New England village was a creation of the nineteenth century, when villages began to serve as commercial centers, not the seventeenth, when settlement remained dispersed. By the mid-nineteenth century the village had assumed a settlement form that was recognizable to Americans.[2]

Although contemporary novelists discussed villages, they rarely described their layout, and prescriptive literature offers the clearest definition of the village ideal. In an 1850 article in *The Horticulturist*, Andrew Jackson Downing recommended the development of "country villages": "a large open space, common, or park, situated in the middle of the village—not less than twenty acres . . . well planted with groups of trees, and kept as a lawn. . . . This park would be the nucleus or *heart of the village* . . . around it should be grouped all the best cottages and residences." Away from the heart of the village were smaller cottages. Throughout the village there would be "sufficient space, view, circulation of air, and broad, well-planted avenues of shade-trees." Similarly, a quarter of a century later, reformer George Waring, Jr., proposed the construction of villages to relieve the isolation of farmers in the West and Midwest. The village of 300 persons would center on an open space with a church, school, and some commercial establishments; the farms would extend out from this core.[3] Thus, the nineteenth-century village in the public imagination had a village green or common, an open space in the center. Plain, two-story houses faced onto the green; a large meeting house would have been the most prominent building. Trees lined the streets. A village was not characterized by a grid of city blocks, extending outward from the core; rather, the village spilled out along its major roads.

The physical similarity of western army forts to this idealized concept of a New England village is striking. Western forts had an open parade ground lined with trees, prominent buildings facing onto that ground, and a rather sloppy extension of outbuildings beyond, consisting of modest dwellings, storehouses, stables, and semi-industrial buildings. Shade trees, bandstands, picket fences, and front porches represented deliberate efforts to create a familiar landscape. Beyond the physical layout, the variety of people who inhabited a fort was reminiscent of a civilian community, including women, children, and other nonmilitary personnel, as well as soldiers—a lively assortment of people that contributed to the buzz of activity. Small wonder, then, that the description most often attached to these forts was "village."

Several factors combined to create the village nature of western forts in the post–Civil War period. One was the increasingly offensive role the army played in the West, rendering massive defensive fortifications unnecessary. Another was the railroad, which altered the army's role while facilitating transportation and supply. Still another was the army's relationship to the profit-driven settlements encouraged by the railroads. And finally, there was the way the forts continued to serve as outposts of eastern American society, representing the Anglo-American cultural establishment that the soldiers were fighting to maintain and extend. The village nature of the forts manifested itself in several ways: in the community of the fort, apparent in the fort's layout and in the people who occupied and visited it; in the buildings themselves; and in the furnishings and conveniences allocated to officers and soldiers.

Purpose

The arrival of the railroad in the late 1860s precipitated fundamental changes in the West. The railroad affected fort-building directly, by altering supply networks, and also indirectly, by causing shifts in government policy toward the Indians. With the railroad also came the products of the industrializing East, including prefabricated building materials and material goods. The new trains also brought settlers. To help finance construction of the vast transcontinental line, the government granted sections of land to the Union Pacific and Central Pacific railroads; the railroads could turn this land into capital only if they could sell it to settlers. The Great Plains evolved from an area to pass through on the way to the West Coast into an area ripe for settlement. The 1868 Fort Laramie Treaty, which allocated certain lands to the Plains Indians, leaving the rest open to settlement, was a product of this changing attitude.

General Sherman recognized the value of the railroad, believing that "every possible encouragement should be given to the Pacific railroad." In his annual report, Sherman mentioned the Union Pacific and Kansas Pacific railroads, then under construction: "When these two great thoroughfares reach the base of the Rocky mountains, and when the Indian title to roam at will over the country lying between them is extinguished, then the solution of this most complicated question of Indian hostilities will be comparatively easy, for this belt of country will naturally fill up with our own people, who will permanently separate the hostile Indians of the north from those of the south, and allow us to direct our military forces on one or the other at pleasure, if thereafter they continue their acts of hostility." With these roads, the army would achieve penetration into hostile territory and could expand its dominance from these thin wedges. Sherman also introduced an important element of his policy, the use of settlers to aid the military mission—in this case serving as a wall between groups of hostile Indians. In addition, settlers helped the army directly, providing needed grain and hay to the army at the frontier posts, enabling the cavalry to range farther.[4]

General Sherman also recognized that the status of the army in the West had changed. Instead of protecting a finite amount of land, and essentially granting the rest to the Indians, the army now controlled a vast expanse of land, with the Indians confined to a specified area. Protection was still necessary, not only for the commercial routes of travel, but also for "the general line of the frontier"—meaning, presumably, the settlements Sherman had been willing to forsake earlier. The army considered any Indian not on his assigned reservation to be "hostile" and subject to military force. Nonetheless, even Sherman recognized that inadequate provisions furnished by the U.S. government often made residence on the reservation untenable, forcing Indians to return to their hunting grounds. All the more reason, Sherman argued, that the government should "force" the Indians to become farmers so that they could provide for themselves.[5]

Confining the Indians to reservations placed the army in the position of defending the Indians from settlers. The most dramatic instance of this turnabout was in the Black Hills of South Dakota, where gold was discovered in 1875. Brig. Gen. George Crook, commanding the Department of the Platte, dutifully reported that he attempted to keep miners and prospectors from infringing on the Indian reservation. The army's effort was often half-hearted. In 1874 Lt. Col. George Custer had led a heavily publicized expedition to determine a site for a new fort, and in the winter of 1875–1876 thousands of miners worked in the Black Hills without interference from the government. Once negotiations to purchase the Black Hills from the Sioux had failed, the government reversed its position and imposed its will on the Sioux. Led by Sitting Bull, who had never signed the 1868 treaty, and Crazy Horse, the Sioux put up a stiff resistance, evidenced by their victories at the Rosebud and Little Big Horn in 1876. But determined pursuit by the army eventually exhausted the remaining Sioux, forcing them to flee to Canada or submit to the reservation in the winter of 1876–1877.[6] Final resistance among Plains Indians ended at the Battle of Wounded Knee in 1890, but by the 1880s the army declared that it controlled the Plains.

The post–Civil War shift in military policy brought an opportunity to change the fort-building strategy. In 1871 Brig. Gen. John Pope, commanding the Department of the Missouri, argued again for concentration and called it a waste of public money to maintain the numerous small posts. His superior, Lt. Gen. Philip Sheridan, commanding the Military Division of the Missouri, responded that it was now the army's duty to give protection to settlers as well as to commercial lines of travel, and that dispersed troops were more effective for these tasks. Renewed warfare in the 1870s prompted the secretary of war, George W. McCrary, to reinforce this policy and even ask for additional forts beyond the 226 that the army then staffed. Finally, the next secretary of war, Alexander Ramsay, argued for concentration in 1880, echoed by Commanding General Sherman. Two years later Sherman issued a list of posts to be abandoned and those to be made permanent.[7]

While the argument over placement of the scattered small forts used the rhetoric of defense, the forts themselves were essentially offensive constructions.[8] First and foremost, the forts were located in "hostile" territory, land that the United States had not yet settled. The presence of these forts was intended to demonstrate government control of the land, and this made their construction an offensive action. Second, the nature of Indian warfare was such that forts were rarely attacked or used in a defensive manner. In the guerrilla war typifying Indian conflict, Indians rarely attacked fixed positions, preferring instead to harass details that had wandered from the greater protection of their more numerous comrades. As a result, when the army sought confrontation, it was invariably beyond the garrisons. The fort provided a base of operations, a source of supplies, and a more permanent residence for the soldiers, but it rarely served as shelter from attack. The design of forts reflected their offensive nature; they tended increasingly to be open collections of buildings without walls, gates, or defensive works.

Third, the army, once perceived as "policemen," became more retaliatory in confronting guerrilla warfare. In the West, individual aggressors easily disappeared into the hinterland. Thus, where the army might have been concerned with tracking individual Indians who had preyed upon immigrants so that offenders could be tried, imprisoned, and punished, instead the army found it more effective to punish whole villages for the depredations of a few of their members. With such actions the army waged a "total war," not limited to warriors but affecting women and children as well. The army's practice of destroying the Indians' food, shelter, and livestock, as well as instances of killing women and children, resulted in a war visited upon aggressors and nonaggressors alike.[9]

The many small forts necessitated by the army's strategy in the West resulted in the wide distribution of the army, which also served another purpose. After the Civil War, the army continued to be an unpopular institution. In the postwar period, Americans were far more concerned with industrialization in the East, with its concomitant problems of immigration and urbanization, and did not worry greatly about the affairs of the Plains. The army was either ignored or condemned by the public at large. For those who feared the tyrannical power of a standing army, the distribution of troops effectively scattered the regiment as a unit of assembly and privileged the company, which was smaller and less threatening. Regiments of 600 men might constitute a force that could harass the citizenry or imperil a democracy, so some thought, but once the regiment was scattered among different posts, the threat decreased. The effectiveness of the fighting force also diminished, but that was not a concern until the close of the century.[10]

Much of the Regular Army's unpopularity in the post–Civil War period related to its Reconstruction duties in the South. In 1867, 40 percent of the army was stationed in the South to secure civil rights for the newly freed slaves. This duty earned the army certain enemies in Congress, who managed to decrease appropriations through the 1870s and withheld pay for the entire force for four and a half months in 1877.[11] Other unpopular duty for the army included its involvement in quelling the strike by railroad workers in several eastern cities in 1877.

The North-South division of opinion on the merits of the military was further complicated by an East-West split on the value of an army in the West. Westerners liked the army, not only for the protection it gave but also for the money it pumped into the local economy. Sales of wood, hay, and produce to army posts, as well as employment for civilian workers, made a military post a boon to a community or even a region. Easterners saw it differently, for to them the West was remote and its needs were not of immediate concern. Historian Russell Weigley has commented on the remoteness of western assignments, as well as an intellectual distance: "The Regular Army was sufficiently isolated to resemble sometimes a monastic order, isolated often physically as it patrolled the distant Indian frontiers, and isolated still more in mind and spirit as it cultivated specialized skills within a sprawling nation of jacks-of-all-trades."[12]

Within the army, morale was low. Due to the number of young officers who had attained high rank during the Civil War, promotion in the dwindling army of the postwar period was exceptionally slow. Although frontier duty provided the only opportunity to see action and thereby attain recognition, it was an unconventional war. Instruction at West Point rarely addressed guerrilla warfare, nor did Emory Upton—a tactical thinker who revised infantry tactics in the 1860s, surveyed armies in Europe and Asia in the 1870s, and proposed reforms for the U.S. Army in the 1880s—concern himself with frontier conditions. The unorthodox warfare, combined with the unsavory aspects of "total war," made frontier fighting a hard sell to an eastern public tempered by the Civil War. Public sentiment was also slowly shifting in favor of the Indians, as reform groups in the East garnered sympathy for their plight. Although the army in the West gained generally favorable press, aided by such officers as George Armstrong Custer who were sensitive to their own public image, the army itself was easily forgotten in the public imagination.[13]

For those involved in the construction of the transcontinental railroad, however, the army was an important partner, and increasingly the army relied on the railroad. The U.S. government provided unprecedented support for this private venture in the form of land and loans. As Sherman saw it, that public investment was enough to justify the use of the army to protect the line. He also recognized that the railroad would aid the military by facilitating the shipment of supplies and the movement of troops. The railroad charged the government half the standard rate; in turn, the government did an immense business with the Union Pacific. For example, the Department of the Platte incurred expenses of nearly $400,000 in 1874 and more than $500,000 in 1876 for shipment of troops and supplies.

Finally, Sherman mentioned that with the railroad, the country "will naturally fill up with our own people," thus achieving the U.S. dominance that he was charged with obtaining. In 1883 Sherman credited the railroad with quelling the Indians; the railroad "used to follow in the rear [but it] now goes forward with the picket-line in the great battle of civilization with barbarism."[14]

In 1866 Sherman established the Department of the Platte, a subdivision of the Military Division of the Missouri, specifically to encompass the Union Pacific Railroad and the Oregon-California Trail. Comprising modern-day Iowa, Nebraska, Wyoming, and Utah, the department was a linear arrangement, reflecting westbound travel. Within this area, the army set out to provide protection for Union Pacific surveying and construction crews. Bvt. Maj. Gen. Grenville Dodge, late of the Volunteer Army but by 1867 chief engineer for the Union Pacific, asked Brig. Gen. Christopher C. Augur for specific troops to accompany his construction crews, as in this note: "Can't part of Stevenson's infantry be put on last 20 miles of work—let's say from point of rocks up above. Let portion guard tie cutters train, then can march with loaded train, and ride back in wagons. We are getting beyond, behind in trees, and cannot get trains to move much without escort. One company to remain in Lawrence Fork, to protect the cutters, is sufficient."[15] Dodge apparently viewed the troops as an arm of his construction team, placing them strategically.

The Pacific Railroad Act of 1862 provided that the subsidy received by the Union Pacific would triple (increasing to $48,000 per mile) once it had passed the "east base of the Rocky Mountains." In June 1867 presidential representative Jacob Blickensderfer, Jr., determined that point to be 525 miles west of Omaha. Dodge located the end of the railroad division nearby, establishing the city of Cheyenne as "the principal depot and repair shops of the company for the eastern base of the Rocky Mountains." Augur, commanding the Department of the Platte, had instructions to locate his fort where Dodge sited the end of his division. On July 4, 1867, Dodge established Cheyenne, and Augur established the military post just 3 miles west of the townsite. He named the new post, like many other army posts, after an officer—in this case, Civil War general David A. Russell.[16] At an elevation of 6,100 feet, the fort sat on the north bank of Crow Creek, occupying a military reservation of 6.8 square miles. The relatively flat site was suitable for railroad facilities but left the fort unprotected, and it was exceptionally windy (fig. 3-2). Cheyenne Depot, a regional distribution station for army supplies, was located halfway between the town and the fort, and was served by a Union Pacific spur line.

Fort and town, just three miles apart, provide an interesting contrast. Fort D. A. Russell, despite an unusual original plan, evolved into a typical

Fig. 3-2. Fort D. A. Russell, ca. 1895. The fort's exposed setting is still evident nearly thirty years after it was established. Courtesy National Archives.

western army post. Cheyenne's history of railroads and cattle and its grid plan made it a typical western town. Despite their intertwined origins, the purposes of the two were very different. In the first few years of Fort D. A. Russell's existence, its troops guarded railroad construction workers and protected the line. Long after construction, the fort's close relationship with the Union Pacific Railroad continued. Troops took advantage of the increased mobility that the railroad provided to participate in confrontations with Indians and other threats to public order. For example, a troop and a half of the 5th Cavalry from Fort D. A. Russell took the train to Fort Fred Steele, near Rawlins, Wyoming, and marched south toward the Ute Reservation in Colorado when hostilities erupted there in 1879. The fort continued to assist the railroad, particularly in 1885 when coalminers employed by the Union Pacific in Rock Springs, Wyoming, strongly objected to the presence of Chinese laborers and rioted against them, killing twenty-eight, injuring fifteen, and destroying company property. The adjutant general used the pretext "for the purpose of protecting United States mails" in order to send federal troops to the scene. Soldiers from Fort D. A. Russell established Camp Pilot Butte in Rock Springs and Camp Medicine Butte in Evanston with buildings constructed by the Union Pacific in an unusual public-private partnership, an inevitable outcome of the close relationship of the railroad and the army. Troops from Fort D. A. Russell rotated in and out of Camp Pilot Butte for the next thirteen years.[17]

Post and Town

The Union Pacific Railroad promised to make Cheyenne an important railroad center, and began selling lots in 1867. By the next spring, however, the railroad began construction of the promised machine shops and roundhouse not in Cheyenne, but in Laramie, 60 miles down the road. Cheyenne's fears for its future were nearly justified when much of the burgeoning town moved on, cutting population from 2,305 to 1,405 in 1870.[18]

Settlements near Forts

Most forts did not spawn cities of the prominence of Cheyenne. Instead, the usual settlement near a fort featured civilians determined to profit from the soldiers through bars, gambling dens, and houses of prostitution. The general character of such communities can be gleaned from Elizabeth Custer's description of Fort Abraham Lincoln's immediate neighborhood: "Just across

> the river from us was a wretched little collection of huts, occupied by outlaws, into which the soldiers were decoyed to drink and gamble... Over their rude cabins they had painted elaborate and romantically expressed signs. In the midst of bleak surroundings rose an untidy canvas-covered cabin, called 'My Lady's Bower,' or over the door of a rough log-hut was a sign of the 'Dew Drop Inn.'"
>
> Similarly, Lt. John G. Bourke described a ride out from Fort Laramie: "Three miles and there was a nest of ranches, Cooney's and Ecoffey's and Wright's, tenanted by as hardened and depraved a set of wretches as could be found on the face of the globe. Each of these establishments was equipped with a rum-mill of the worst kind and each contained from three to half a dozen Cyprians [prostitutes]... who lured to destruction the soldiers of the garrison." Settlements with bars and prostitutes, as well as housing for civilian workers, did not compete with the fort for architectural excellence or household conveniences, nor did they offer the social life to which officers aspired. Instead, they offered an example of what high-minded officers and their wives aimed to avoid in the kind of civilized life they established at their forts.
>
> Sources: Custer, *"Boots and Saddles,"* 229–230. Bourke cited in Hedren, *Fort Laramie in 1876*, 46.

Union Pacific engineers laid out lots in Cheyenne on a grid plan, with no parks or public squares, maximizing the value of the land (fig. 3-3). As building materials were scarce, Cheyenne consisted of tents and shanties. Several buildings arrived by rail from Julesburg, the previous end of the line, in a pattern that had repeated itself as the railroad moved along. One observer described the arrival of a freight train in Cheyenne, "laden with frame houses, boards, furniture, polings, old tents, and all the rubbish which makes up one of these mushrooming 'cities.' The guard jumped off his van, and seeing some friends on the platform, called out with a flourish, 'Gentlemen, here's Julesburg.'" The population that followed the road was reportedly a shady one, full of gamblers and prostitutes, not the sort of people who were interested in establishing a civilized society.[19]

One of the fort's primary roles in Cheyenne was to protect the railroad's interests as much as those of citizens. The army maintained order and prevented squatting. By 1869, though, the city was imposing order on the soldiers. Cheyenne police habitually arrested soldiers who were in town on payday, perhaps for good cause, although the soldiers complained that they were treated unfairly. When a controversy arose over one of these arrests, the fort prohibited all soldiers as well as officers from visiting the city, and Cheyenne immediately felt the economic impact. The *Chicago*

Fig. 3-3. Bird's eye view of Cheyenne, 1870. The Union Pacific Railroad laid out the new town of Cheyenne in a grid plan. In the left foreground is the railroad roundhouse. In the distance is the army's quartermaster depot, and beyond it, Fort D. A. Russell in an unrealistic configuration. Courtesy Wyoming State Archives, Department of State Parks and Cultural Resources.

Tribune noted, "The soldiers have brought large sums of money into the city, and spent it freely, to the great advantage of the tradesmen and merchants." The fort's payday amounted to $52,000. While soldiers bought liquor in the city, officers' wives purchased all types of domestic goods. Elizabeth Burt, an officer's wife, noted in 1869 that "Cheyenne had grown rapidly into quite a flourishing town, supplying some of the luxuries in addition to the actual necessities of life but at prices calculated to impoverish an officer." She mentioned a few such luxuries: "Corn on the cob was a great treat, especially as it was the first we had eaten in three years. Peaches and pears from California were enjoyed sparingly as eating them was like consuming gold. Luscious red watermelon too, cold and tempting, how delicious!" In these early years, the fort was more sophisticated than the town. Cheyenne citizens with social pretensions could hobnob with officers who had social graces and cosmopolitan airs.[20]

The fort also provided a boost to Cheyenne's economy by hiring numbers of civilians, especially for freighting. The establishment of the Cheyenne Depot between the fort and the town meant that this area would serve as a supply base. The army sent goods as far as it could by rail—to Cheyenne—and then freighted them overland to posts such as Fort Laramie and Fort Fetterman. The army employed hundreds of freighters to transport goods to remote posts. The fort also boosted the local economy through the personal investments of its officers. As the local newspaper often mentioned, Col. John D. Stevenson, Fort D. A. Russell's first commander, invested in Cheyenne real estate. Besides purchasing a corner lot, Stevenson "also built a vast warehouse in Cheyenne, a regular warehouse in stone, if you please, and not of wood." Other officers invested in cattle; in 1883 the *Cheyenne Daily Leader* named six officers who had livestock holdings.[21]

Cheyenne boomed during the Black Hills gold rush, when nearly one-fourth of the traffic to the gold mines in 1876 and 1877 passed through the town. Cheyenne's newspaper greeted the iron bridge constructed over the North Platte River near Fort Laramie as "the beginning of a new era in the development of Cheyenne." This crossing of the Platte meant that one of the best routes to the Black Hills was to take the railroad to Cheyenne, and the Cheyenne & Black Hills Stage, Mail, & Express Line north from there. New hotels, banks, warehouses, and people appeared in town, and Cheyenne's future was secured.[22]

The U.S. government's treaty with the Sioux guaranteed their access to the Black Hills and northeastern Wyoming. When the government abrogated that treaty, the Black Hills were opened to gold miners and northeastern Wyoming to cattle ranchers. Cheyenne did not become a "cow town" like Dodge City, Kansas, through which cattle were shipped, but it did become the home of some immensely wealthy ranch owners. The fifty-one stock growers listed in the 1880 census included men such as Alexander Swan, who owned more than forty-five thousand head of cattle in eastern Wyoming, assessed at more than $150,000. Forty percent of the members of the Wyoming Stock Growers Association listed Cheyenne as their residence. By 1883 Cheyenne's newspaper touted its wealth: "Cheyenne is said to be the richest city in America, the population taken into consideration. The city has a population of about 5,000, and the residents are worth at the least possible calculation $50 million." A slightly more sober report from the *Laramie Sentinel* gave different figures but drew the same conclusion: "It is computed from the census and assessment rolls that the wealth of the city aggregates $4,640 per head to every man, woman, and child in the city. . . . We are willing to concede that it is the wealthiest city of its size on the globe."[23]

With this wealth, Cheyenne received several votes of confidence, allowing it to take on the trappings of civilization. The Union Pacific Railroad finally built its repair shops in Cheyenne after 1875. In the 1880s it moved its transport department there, built an extensive system of shops, and constructed a handsome new passenger depot. The territorial government, which had been located in Cheyenne since its establishment in 1868, appropriated $150,000 to build a new capitol there in 1886 (fig. 3-4). In 1890 Wyoming became a state. The fort, too, experienced reinvestment in the 1880s. In 1882, when General Sherman decided that Fort D. A. Russell should be rebuilt into a permanent institution owing to its location on the railroad, the newspaper wistfully expressed its hope that the fort would move closer to town "to enable the garrison to reap the benefits of town life" and, undoubtedly, to enable the town to reap the benefits of the soldiers' paychecks. The fort did not move, but it did undergo extensive new construction. Although the new buildings were brick and much more substantial than the previous barracks and quarters, they did not begin to rival the ostentatious cattle barons' homes in Cheyenne. Eclectic in style, these houses displayed their owners' wealth in a profusion of porches, towers, turrets, bay windows, and rich materials (fig. 3-5). Carey Avenue, then called Ferguson Street, was known as "Millionaires' Row."[24]

Against this backdrop of a fast-developing Western city celebrating new wealth, brash cattle barons, and up-to-date conveniences transported on the railroad, a fort such as D. A. Russell may have appeared old-fashioned and slow-paced. After its first year or two, when it helped establish order among white settlers, the fort settled into an almost peaceful existence, with its band concerts drawing the most notice from the local newspaper. With a shabby elegance, the fort reflected old class lines articulated as ranks, not the nouveau riche excitement of a boom town. This contrast may have been one of the reasons why the fort was described as a "village."

Novelists such as Henry Ward Beecher, Harriet Beecher Stowe, Mary E. Wilkins Freeman, and Sarah Orne Jewett documented the ethos of the village in regional literature, idealizing the old-fashioned values and sense of community found among hard-bitten New Englanders in picturesque settlements.[25] To them, the village connoted hearth and home, domestic environments of simple, unpretentious people unaffected by the momentous changes wrought by industrialization and urbanization. Villages were not without class lines; most of the households featured in the literature had servants. But in a village everyone knew their place, just as in a fort everyone knew their rank. In a western boom town, such certainties were shattered.

Layout

The design of Fort D. A. Russell initially had a clear conception. Fortifications were deemed unnecessary, as the presence of the railroad and the

Fig. 3-4. Bird's eye view of Cheyenne, 1882. By this time Cheyenne was a fully developed city. The Union Pacific donated four city blocks to form City Park, shown here as fully landscaped. Courtesy Wyoming State Archives, Department of State Parks and Cultural Resources.

Fig. 3-5. Nagel-Warren house, 222 E. 17th St., Cheyenne, photograph by J. E. Stimson, ca. 1900. Cattle barons built ostentatious houses in Cheyenne, such as this one built in 1888 for Erasmus Nagle, and later occupied by Sen. Francis E. Warren. Courtesy Wyoming State Archives, Department of State Parks and Cultural Resources.

tent city of Cheyenne made the fort's surroundings settled from the start. Unique among western army posts, the plan of the fort was in the form of a diamond, with the buildings arranged *en échelon* (fig. 3-6). In this arrangement, barracks faced the parade ground obliquely, not occupying the full frontage of the parade ground, so that more buildings could be accommodated along its perimeter. As explained by the post surgeon, Charles H. Alden, who claimed credit for the design along with the commanding officer, Col. J. D. Stevenson, "The diamond form of the parade was adopted not only for the sake of appearance, but to avoid the inconvenience of the very large inclosed space, which would have resulted from the ordinary rectangular or square space, owing to the great number of buildings required." Alden wrote somewhat defensively of the plan: "The plan has been sometimes objected to as wanting in the regularity usually found and by many deemed essential in a military post. The special object of the plan was to group the many buildings so as to bring them within a convenient distance of each other and yet not crowd them and cut off in any way the access of light and air."[26] Alden's reference to the "regularity" of other posts recognizes the ubiquity of rectilinear parade grounds and conventional arrangement of primary buildings, not a rigid standardization. And despite Fort D. A. Russell's idiosyncratic plan, it did observe the conventions of an open parade ground surrounded by barracks and quarters.

Fort D. A. Russell's plan had a clear hierarchical arrangement. At the north point of the 800-by-1,040-foot diamond was the commanding officer's quarters; on the northwest side of the diamond were the infantry officers' quarters; the cavalry officers' quarters occupied the northeast side. At the south point of the diamond was the guard house; on the southwest side were the infantry barracks; the cavalry barracks occupied the southeast side. All other buildings—stables, workshops, warehouses, and more— were behind these, off the parade ground. There was also a high degree of architectural unity: all buildings were wood frame clad with board-and-batten siding, with essentially two building forms—officers' quarters and barracks.

The coherent design of a fort such as D. A. Russell was difficult to maintain, however. Different commanding officers and post quartermasters oversaw the construction of new buildings, locating and designing them according to their own imperatives. New building forms also appeared, especially off the parade ground. Comparison of plans of Fort D. A. Russell in 1870 and 1875 illustrates this growth. On the 1870 plan, buildings off the parade ground included the following: "Brown's Mess House," a restaurant for officers; the hospital, behind which were the sink (latrine), dead house (mortuary), and rain gauge, and in front of which were quarters for the surgeon and assistant surgeon; the post trader's store and house; a semi-industrial area including the carpenter shop, quartermaster's quarters, commissary storehouse, quartermaster's storehouse, butcher shop, bakery, and wood yard; quarters for the ordnance sergeant; ice house; and steam engine. Five years later this community had grown with the addition of the following: forty small buildings identified as laundresses' quarters; separate quarters for the band and the post guides; tailor shop; school and library; butcher's, tin, and saddler's shops; coal house; stables; blacksmith shops; and sawmill.

The addition of secondary buildings off the parade ground could continue indefinitely, but the limitations of the parade ground configuration proved themselves when the addition of barracks and officers' quarters became necessary. A parade ground with buildings on four sides could not be expanded; instead, the primary buildings started bleeding out the edges. When new brick captains' quarters were added to Fort D. A. Russell

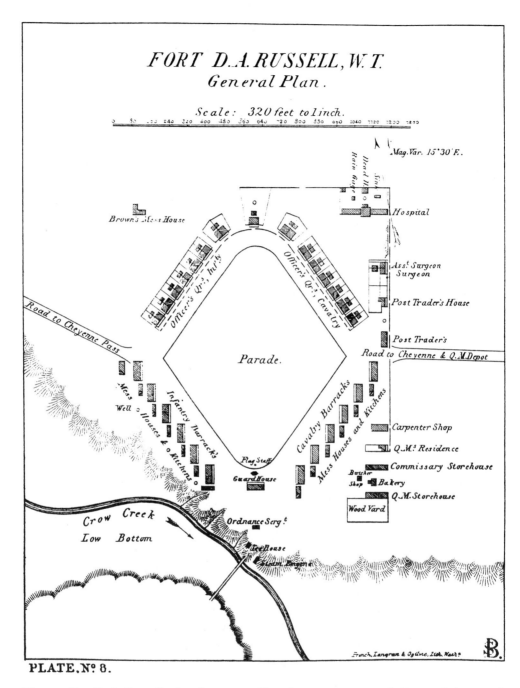

Fig. 3-6. Fort D. A. Russell, site plan, 1870. The plan was an unusual diamond shape and the barracks were arranged at an angle, enabling them all to face onto the parade ground without being packed too tightly. From U.S. War Department, Surgeon-General's Office, Circular No. 4, *A Report on Barracks and Hospitals, with Descriptions of Military Posts*, pl. 8.

Fig. 3-7. Fort D. A. Russell, 1922. The original diamond-shaped parade ground is in the foreground, but expansion has resulted in lines of officers' quarters, barracks, and stables stretching eastward toward Cheyenne. Courtesy Wyoming State Archives, Department of State Parks and Cultural Resources.

in 1885, some were placed on the parade to replace quarters that had burned down. But the remainder faced onto the road to Cheyenne, which headed off the east point of the diamond. As new construction continued in the twentieth century, it spread in the same direction, with quarters on the north side and barracks on the south (fig. 3-7). Today the original diamond is almost lost in the extended polygonal parade ground.

The forts' tight arrangement of primary buildings facing onto an open space inhibited rational expansion. Additions to settlements were best accommodated by the most common town arrangement in the West, the grid plan, which was employed in Cheyenne. This regular, rectilinear arrangement of city blocks could be expanded across the Plains indefinitely, and was particularly suited to the entrepreneurial activity of town promotion, as it produced many lots of theoretically equal value, with little wasted, unmarketable space. But the grid plan was ill suited to forts. Aside from practical and ceremonial problems of troop assembly, the grid plan's association with urban centers of the East and the marketing of the West was at odds with the army's re-creation of an atmosphere of tranquil domesticity and civilized refuge evoked by the village model.

Other forts experienced the same kind of expansion as Fort D. A. Russell. Although the railroad bypassed Forts Laramie and Bridger, they continued to be active posts until 1890. Despite the cessation of much of the westbound traffic on the Oregon-California Trail after completion of the transcontinental railroad in 1869, Fort Laramie continued to be an important troop assembly point for the Plains Wars of the 1870s; the fort saw its greatest amount of activity in 1876. The assignment of an additional two companies in the summer of 1872 necessitated new barracks. With no room to spare, the new troops took over a building that had just been completed across the river from the post for the use of post laundresses. A request for construction of new barracks went forward in 1873; construction began the next year.[27] Since there was no room on the parade ground for the new structure, it was placed in line with one side of the parade ground, but far beyond its limits.

Expansion of the posts in odd directions off the parade ground with a mix of building types and materials contributed to the image of these forts as villages. A village might well have a mix of buildings and a sense of organic, unplanned growth, as these posts did. General Sherman described Fort Laramie in 1866 as "a mixture of all sorts of houses of every conceivable pattern, and promiscuously scattered about."[28] While Sherman was criticizing the unmilitary appearance of such a collection of buildings, he was also describing a quality that made the forts appear as villages.

The proximity of the officers' quarters to each other on the parade ground also fostered a sense of community much like a village. Officers' wives, indefatigable recorders of army life, certainly found that quality of village life. Cynthia Capron described life in Fort Laramie in 1876: "In mild weather people almost lived on their porches, and generally, the houses were in those days built around a square—the parade ground—each house being in full view of the others. There was a nearness and a feeling of sociability."[29]

Sometimes neighbors seemed too near. The placement of lower ranking officers in double quarters was a constant source of complaint. It was an obvious economy move, as the shared wall made double quarters cheaper than two singles. But in an area of wide open space, officers resented being located so close to each other. Lt. Col. Thomas M. Anderson pointedly wrote, "When we read of double houses and blocks of residences we are compelled to wonder if the land is selling by the foot in that remote region." While Anderson mentioned the fire hazard, it was probably the proximity of neighbors that most offended occupants of double houses. With the introduction of piped gas and water in the 1880s and 1890s, new reasons emerged for locating officers' quarters close to each other. As in a city, provision of utilities was less expensive in a dense area. At this time, though, a shift in military policy produced brick officers' quarters on a grander scale than before, making double quarters even less desirable.[30]

While the domestic nature of officers' quarters was one of the most village-like components of the fort, some building types were missing. New England villages were characterized by meetinghouses, but one building type that was conspicuously absent from these three forts was a chapel. The army authorized chaplains beginning in 1838, but they were not present at every post. Generally, religion was downplayed at western posts and, consequently, so were possible religious tensions. In a money-tight environment, post quartermasters experienced difficulty in getting approval for construction of chapels, so chaplains improvised, holding services in any available building. One ward of the hospital served as a chapel at Fort D. A. Russell in 1870 and 1875, but by 1879 chaplain Jeremiah Porter was holding services in a small room in the headquarters building, until attendance was so good that his congregants could no longer fit. In 1885 at Fort Laramie the chapel shared a former barracks with the fire engine; that year the fort received a new concrete administration building with one wing devoted to a chapel with multiple uses: "a fine chapel convertible into a 'hall,' where we listen to the music of our excellent band or witness the equally enjoyable performances of our amateur theatrical people" (fig. 3-8).[31] The multipurpose building did not have the appearance of a chapel, though, so, like Fort D. A. Russell, the post remained visually secular.

Libraries were another building type found in the New England village landscape but absent from most forts. The army authorized reading rooms, beginning in 1866, but these were usually located in buildings devoted to other functions. At Fort Laramie in 1870 the library was located in the adjutant's office, a separate building near the two-story officers' quarters. In 1883 the reading rooms shared quarters with bathing rooms. Two years later the "library and reading room" occupied its own building, a frame barracks measuring 20 by 80 feet that had been built at least twenty years before. The various reading rooms may not have made an impact architecturally, but they did represent a literate society. Besides a collection of several hundred volumes, the post library held magazines and newspapers. In 1889 Fort Bridger subscribed to five daily papers, four magazines, and two military journals.[32]

Landscaping

The landscape was another aspect of forts that evoked villages. The absence of trees on the Great Plains was the starkest reminder that military

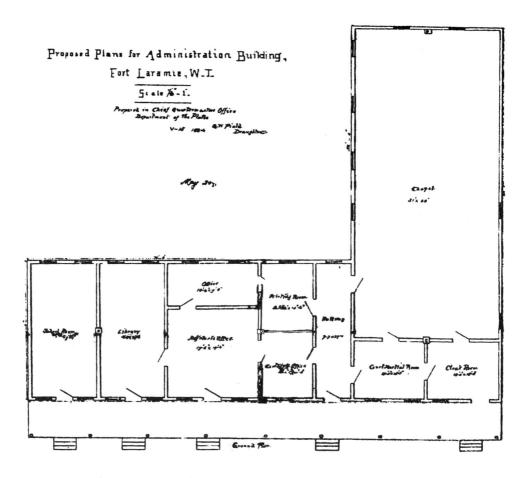

Fig. 3-8. Fort Laramie, administration building, plan and elevation, 1884. Chapels rarely received their own buildings. At Fort Laramie the chapel doubled as a theater and was incorporated into the administration building, which was built in mirror image of this proposed plan. Courtesy National Archives.

families were not in the East any more. Beginning with the post–Civil War regarrisoning of western posts, post commanders oversaw the planting of trees, as perhaps the strongest indication that the army was building more than military posts, and was venturing into the civilizing process. In 1870, just three years after its establishment, Fort D. A. Russell received cottonwood and pine trees, which were planted around the parade ground as well as off it (fig. 3-9). Fifteen years later, in the midst of a considerable building project, the post quartermaster received permission to buy a thousand shade trees, and had to ask which account to bill them to. The secretary of war ruled "that shade trees were a shelter to a garrison," so he was permitted to draw on the barracks and quarters appropriation. Twenty years after that, however, a requisition for 750 feet of garden hose and twelve lawn sprinklers was disallowed. At Fort Laramie in 1872 a lieutenant took a detail of ten men to procure fast-growing cottonwood trees for the parade ground. Photographs indicate that by the 1880s Fort Laramie's parade ground was lined with trees (fig. 3-10). Although Fort Bridger was favored with quite a few trees, troops had less success with transplanting them, as the post surgeon observed: "Trees and shrubbery were planted on the parade; but most of them have died, and the others have had a very slight growth, thus corroborating the experience of other officers who have transplanted trees from the vicinity without success."[33] Cultivating trees at these forts required some tenacity, as they needed irrigation.

The treatment of the parade ground, analogous to the village green, was important. To permit the troops to parade, the space had to be unimpeded, except for a bandstand and flagpole. Trees might line the perimeter, but could never be allowed to invade the center. At most posts the parade ground was bare dirt; as Elizabeth Burt noted of Fort D. A. Russell in 1868, "the wind constantly [swept] the parade ground bare." If it was not bare dirt, the parade ground had to be mowed. Mowing machines were required at posts to cut hay, not solely to manicure the parade ground. Margaret Carrington described the process of creating a parade ground at the new Fort Phil Kearny: "To secure at the outset a handsome and permanent parade-ground the long train of wagons was repeatedly driven about the designated rectangle . . . while a mowing machine soon cut the grass and gave the start to the present beautiful lawn of the Fort Phil Kearney Plaza."[34]

One construction permitted on the parade ground was a bandstand, which appeared at posts accommodating the regimental band (fig. 3-11). A bandstand was consistent with Downing's view of a country village: "And, little by little, the musical taste of the village (with the help of those good musical folks—the German emigrants) would organize itself into a band, which would occasionally delight the ears of all frequenters of the park with popular airs." The lawn of Fort Phil Kearny had a bandstand: "A flag-staff,

Fig. 3-9. Fort D. A. Russell, officers' quarters, early 1870s. Soon after the fort was constructed, the army planted cottonwood trees on the parade ground in front of the officers' quarters. Courtesy Wyoming State Archives, Department of State Parks and Cultural Resources.

surrounded by an octagonal band platform, stand, and seats, occupies the center of the parade." Fort Bridger gained a bandstand in the late 1870s, which complemented the picket fences, footbridges over the creek, and numerous trees to create a pastoral setting.[35]

Each fort had a post garden, located well off the parade ground, to produce vegetables for the soldiers. Although post surgeons saw gardens as essential for the prevention of scurvy, difficulties in irrigating a garden in a semi-arid climate, particularly in summer when most men were out on patrol, meant that post gardens' yields were disappointing. At Fort Laramie in 1850, Assistant Quartermaster Stewart Van Vliet sent an agent to New Mexico to hire Mexicans to tend the post garden. He explained that "it is only by irrigation that the land in this vicinity can be cultivated & I have no one who understands it." Mexicans, he thought, would bring the proper expertise to farming in a semi-arid climate. The Mexicans, if they were ever actually brought in, did not make much of an impact, and soldiers continued to struggle with irrigation. Grasshoppers also plagued Wyoming post gardens in the late 1860s. Hospital gardens received special care, not only as vegetable gardens to provide patients with important nutrients, but also as flower gardens to aid the recuperative process (fig. 3-12). The post surgeon noted the planting of "a large number of flower bulbs" behind the

A BEAUTIFUL VILLAGE ■ 103

Fig. 3-10. Fort Laramie, officers' quarters, 1889. By the 1880s, cottonwood trees were thriving on the parade ground at Fort Laramie. Courtesy National Archives.

hospital in 1869, and an even greater effort was expended on the yard of the new hospital a few years later.[36]

Individuals, particularly officers' wives, contributed to the greening of the forts in a more ornamental way. As George Forsyth reported: "If the post is one of a few years' standing, it is safe to say that the dooryards and porches of the officers' quarters, and frequently the barracks of the enlisted men, will be embowered in vines and flowers. It is a rare exception when the wives of military men at frontier posts are not fond of trees and flowers, and do not spend a few moments each day during the summer season in personally caring for them, with the result that garrisons frequently present a very homelike and restful appearance." Plantings were not solely a feminine task, however. One officer suggested that if each officer planted an ornamental shrub, then when he changed posts he would find a similar planting at his new quarters. Every officer was advised to plant in front of his quarters "a berry-bearing bush, a rose tree, a grape vine or perhaps a young shade tree."[37]

Fig. 3-11. Fort Meade, South Dakota, trumpeters of 8th Cavalry at bandstand, 1894. A bandstand was the one construction permitted on the parade ground, as music helped unify marching soldiers. Courtesy Library of Congress.

Officers' quarters had yards, although not always defined by fences. An officer's complaint about the proximity of buildings illustrates the life that went on in these yards: "The houses were built on a large reservation, but crowded so close together as to be a nuisance to the occupants, bringing the tenants into most uncomfortable propinquity, which was further aggravated by the total absence of any division fences to the yards in the rear, so that there is a free range for servants, peddlers, dogs, and the vagrant public, among clothes lines, hen coops, *et id genus omnia*, to visit the kitchens promiscuously without hindrance, for tittle tattle and gossip broadcast." Officers particularly desired fences if they occupied double houses, as did Lt. Henry E. Robinson at Fort Bridger in 1881, who requested a "board fence in rear of his quarters dividing back yard if quarters next to his are to be occupied." Even if the back yards had fences, they were not always good ones; one visitor to Fort Laramie in 1876 described "the unpicturesqueness of the yards, the coal- and wood-sheds, the rough, unpainted board

Fig. 3-12. Fort Laramie, hospital garden, 1888. The hospital's porch, picket fence, and flower garden created a domestic setting for patients. In this view, Assistant Surgeon Louis Brechemin is at the left, next to the hospital steward. Courtesy the Brechemin Collection, Fort Laramie National Historic Site.

Fig. 3-13. Fort Laramie, officers' quarters, late 1880s. Shade trees, picket fences, boardwalks, and street lamps create a village-like setting for the mansard-roofed commanding officer's quarters. Courtesy American Heritage Center, University of Wyoming.

fences; the dismantled gate, propped in a most inebriate style against its bark-covered post."[38]

The front yards of the quarters were more formal, with picket fences imparting a domestic character. At Fort Laramie, picket fences lined boardwalks that ran in front of the officers' quarters; the installation of mineral-oil streetlamps in 1883 further contributed to the village atmosphere (fig. 3-13). At Fort Bridger in the 1880s, picket fences defined the officers' quarters' front yards (fig. 3-14). A boardwalk ran in front of the fences, parallel to a dirt road. The road was flanked by a row of trees, irrigated by acequias, or ditches. The creation of home was evident in the treatment of the quarters' yards, just as the creation of a village was reflected in the walks and roads that linked them.

Some parts of the landscape were devoted to the same group activities that took place in other American villages. Fort D. A. Russell had a baseball diamond by 1870 and a bowling alley. Seeing value in recreational opportunities, Quartermaster General Meigs encouraged men to build bowling alleys with their own money, as appropriations were tight. Journalist Richard Harding Davis observed class distinctions in the recreational landscape; he noted that the typical post had a baseball diamond marked out

Fig. 3-14. Fort Bridger, 1890. Landscaping in front of the officers' quarters, off to the left of this photograph, included picket fence, boardwalk, irrigation ditches, and rows of trees. The tricycle parked in the left foreground is likewise unmilitary. Courtesy National Archives.

near the barracks, and tennis courts near the officers' quarters.[39] A landscape feature such as a baseball diamond is easily overlooked unless it is filled with athletes, cheered on by an audience. Understanding the liveliness of these village-like landscapes involves a brief examination of the kinds of people living, working, and visiting there.

Community

The landscape was not separate from the people who built, tended, and inhabited it. Soldiers were not the only people at a fort: women, children, workmen, traders, and even Indians lived there. Some of them had purpose-built buildings, such as the post trader's store, an integral part of every fort. Others remained invisible in the landscape, such as officers' wives who resided in whatever quarters their husband's rank entitled them to. But all these people contributed to the life of the place, wearing different-colored clothing, speaking in different accents, walking with different bearings, using the landscape of the fort in different ways. Like a village—and surely one of the reasons observers described forts as such—the fort had a mix of people of different ages and genders performing a multitude of tasks.

If the sounds of the fort were as easy to re-create as the sights, the variety of people would be more obvious. The soldiers themselves spoke in various accents, many from outside the United States. Of 268 privates at Fort Laramie in 1860, 169 (63 percent) were foreign-born. In 1880, 155 (43 percent) of the 363 military personnel at Fort Laramie were foreign-born. Of these, 62 were from Ireland and 46 from present-day Germany. Similarly, of the 54 recruits enlisted at Fort D. A. Russell in 1883, 19 (35 percent) were foreign-born; 9 of these were from Germany, 7 from Ireland. These percentages of foreign-born soldiers were consistent with nationwide trends. In addition, after the Civil War, companies of black soldiers, who constituted about 9 percent of the strength of the army, were stationed in the West. Although segregated in their own companies, and therefore inhabiting their own barracks, African Americans were posted at mixed garrisons with white companies.[40] None of these African-American companies, though, was posted at any of the three forts under discussion here before 1890.

Officers differentiated themselves from soldiers on many levels. Obviously, officers had commissions, received the privileges of their rank, and performed different tasks than enlisted men (fig. 3-15). They were also better educated and tended to be native-born; of the twelve officers at Fort Laramie in 1880, none was foreign-born. The pay scale reinforced the difference between officers and enlisted men. From 1871 until the end of the century, the base pay of privates was $13 per month, plus rations; sergeants earned from $17 to $34 per month. By contrast, officers earned more than $100 per month, without rations. Second lieutenants received $1,400 per year, and first lieutenants, $1,500; pay increased with rank up to $3,500 per year for colonels.[41]

Officers perceived themselves as being of a different, higher class than enlisted men. Although ranks might have created levels within this upper class, in fact the officers socialized together. Because a lowly lieutenant had the potential to become a colonel or a general, he was a member of the same class; it was much rarer for an enlisted man to become an officer. The hierarchical form of military housing reinforced the class difference. Soldiers and noncommissioned officers lived on one side of the parade ground, in large barracks. Officers lived on the other, in quarters that resembled houses. Although these quarters were assigned according to a strict hierarchy, with a higher ranking officer receiving larger quarters, a commanding officer's quarters might not actually look any larger than the double houses occupied by two captains. The architecture thus reinforced the class divisions present at the fort: a vast gulf between officers and enlisted men, and subtle differences among the officers.

Fig. 3-15. Fort Bridger, probably 1870s. Officers and their families gather for a game of croquet on the parade ground in front of the officers' quarters. Post trader William A. Carter is the bearded man seated on the right. Courtesy Wyoming State Archives, Department of State Parks and Cultural Resources.

The women at an army post also had a strict class orientation, deriving from the men with whom they were associated. Officers' wives were invariably referred to as "ladies"; the wives of enlisted men, and any working women at the post, were called "women." The only females whom the army officially accommodated were laundresses (fig. 3-16). The army assigned laundresses to each company at a rate of one per nineteen men. Laundresses received rations and quarters, but for income they charged for their work. At Fort D. A. Russell in 1868 their rates were set as follows: soldiers' laundry, 75 cents per month; officers' laundry, $3.00 per month; and officers' wives' laundry, $4.00 per month. Laundresses were usually wives of enlisted men or noncommissioned officers; in an official list of the twenty-three laundresses at Fort D. A. Russell in 1868, five were apparently unattached, as they were not given the title "Mrs.," and two were apparently unmarried daughters of other laundresses. With a reputation as hardened, low-class women, laundresses were often the object of amusement for outsiders, as illustrated by an item that appeared in the Cheyenne newspaper in 1883: "Quite a ludicrous scene was witnessed here Tuesday afternoon in front of the laundresses' quarters, better known by the sobriquet of 'soap suds row,' in a fight between two pet dogs . . . their

Fig. 3-16. Laundress, with Pennsylvania 31st Infantry, near Washington, D.C., 1862. The army appointed laundresses, granting them rations and quarters. Unlike officers' wives, they had an official role at the post. Courtesy Library of Congress.

dogships were going in heavy when their respective owners appeared upon the scene, newly arisen from the 'army piano' (the washboard), and rushing frantically to the rescue of their pets, expended 'volley after volley' of blank—'Ye Gods!' what a sight!. . . amidst the yelling of the spectators, the yelping of the dogs and the screaming of the children, the first note of the bugle sounded for the sham battle." The newspaper article repeated common prejudices about laundresses: they were profane, fearless, strong, and unfeminine. In other venues, laundresses were also praised as industrious and kindhearted. The reality, of course, was more complex. In 1878 the army discontinued its support of laundresses, reasoning that the expense of housing and moving them was too great. They were permitted to stay on until their husbands' five-year enlistments ended, but by 1883 army laundresses were no longer officially on the rolls.[42]

The army actively discouraged soldiers from marrying by having them attest at their time of enlistment that they had "neither wife nor child." Once enlisted, they could marry only with the permission of their company commanders, and married men were permitted to reenlist. Enlisted men and their wives often occupied minimal housing on the post, perhaps built at the expense of the residents themselves. Soldiers' wives remained obscure in the written record, but officers' wives sometimes mentioned them as servants or as aiding in childbirth. Orders banning single women and enlisted men's wives from the post indicated that they must have been there, occupying found housing. The same order issued at Fort D. A. Russell in 1868 that listed all the laundresses by name and the amount they could charge for laundry also banished all other women: "All other females in the garrison, except those living with officers' families, Mr. Brown or the Post Trader, will leave in 24 hours." Prostitutes operated quite openly, but off the post, outside the military reservation.[43]

Officers' wives had a presence greater than their actual numbers, due to several factors. For one thing, they published their reminiscences; such works by army wives in the West abound. They often had a specific agenda, such as Frances C. Carrington's attempt to clear her husband's name after the Fort Phil Kearny massacre, or Elizabeth Custer's in keeping the flame of her husband's popularity alive. Others took a romantic view of the wilderness, the danger, the bravery, and their own civilizing role and the good they were doing their husbands by being there. As Julie Roy Jeffrey found in her study of frontier women, army women clung tenaciously to their roles as guardians of the home and purveyors of culture.[44] Their accounts highlighted their efforts to domesticate their quarters, provide cultured entertainment, and bear up under adversity without complaint.

It is curious in some respects that officers' wives were allowed in what was essentially a war zone, but Frances Carrington contended that General Sherman himself had encouraged "all army officers' wives to accompany their husbands and to take with them all needed comforts for a pleasant garrison life in the newly opened country." Margaret Carrington, the colonel's second wife, who as the wife of a lieutenant went on the same expedition as Frances, remembered Sherman's advice as "preserving a daily record of the events of a peculiarly eventful journey," undoubtedly to sway public opinion.[45] Sherman's strategy was probably twofold: that women would bring eastern civilization to the West, thus achieving the army's overall purpose, and that women would write letters home or publish books that would enhance the image of the army.

Samuel B. Holabird, a few months before he was appointed quartermaster general in 1883, articulated the role of officers' wives at western posts: "The generous and beneficent influences of woman are nowhere more apparent than at the remote Army posts on the frontier. She is a constant balance wheel to the social machinery, with an influence always on the side of good order, kind and neighborly feeling and refined intercourse." The room allocations that severely restricted the number of rooms to which an officer was entitled appeared to discourage marriage. One officer encouraged the army "to acknowledge the fact that as long as there is no law or regulation forbidding an officer to marry he will follow the dictates of nature in this matter, and to provide for the contingency in the way of more commodious quarters."[46]

Holabird perceived a direct connection between substantial architecture and an improved moral tone, because better buildings would permit and encourage officers to bring their wives with them to remote posts. "Posts wisely built will naturally elevate the character of the service with both officers and men, for it will render it both easy and proper for them to take their families to the scene of their duties, thus securing to them the restraining and refining influence of society with greatly added incentives to duty; a feeling of contentment, not usual, with their lot, and besides the public advantage of a colonizing tendency to strengthen the Territory where the post is situated. Innumerable moral and physical advantages would be sure to follow a good system of quarters."[47] Holabird cited not only the wives' benevolent influence on their husbands, but also the "public advantage" of having wives "colonize" the territories in which they were posted. He apparently believed that women's benevolent influence would extend beyond the post, to its surroundings, or perhaps he was hoping that military families would choose to settle in these regions when their tour of duty was completed. Nonetheless, the sentiment in favor of officers' wives did not extend to the wives of enlisted men. The latter were rarely credited with middle-class virtues of civilizing and culture-mongering.

Despite the perceived benefits that officers' ladies would bring to army posts, official regulations provided nothing for them. As Elizabeth Burt explained it, "'What about the ladies?' do you ask? The Army Regulations provide nothing for wives and children, not even mentioning their existence." Officers' wives received no rations, moving allowance, or quarters other than what their husbands were entitled to, a source of constant aggravation in the women's reminiscences. But mostly, the women behaved as Sherman thought they would, by generating sympathetic accounts of the western army and by undertaking their civilizing duties. Officers' wives entertained each other, or urban neighbors if there was a town nearby; they put on theatricals and attended musical performances by the regimental band; they held dances and masked balls. They instructed and deplored their servants, they fretted about how to furnish a house or how to compose a meal from limited resources. They had babies, taught children, and kept familial ties to their homes in the East.[48]

Elizabeth Burt, who accompanied her husband to various posts in the West, including Forts Laramie, Bridger, and D. A. Russell, penned reminiscences characterized by gritty determination, such as the following passage, describing her arrival at the commanding officer's quarters at Fort Bridger in 1866: "Three months of tent life had not made us critical of our new surroundings, although the post was not in particularly good repair and no quartermaster supplies were to arrive before the coming year. The only thing for us to do was to look on the bright side and be thankful that we were all together, in good health and sheltered in a substantial home." Burt, who brought her sister with her for female companionship, rejoiced at finding other officers' wives at each post and carefully enumerated them as a way of defining her social circle. In the post–Civil War period, there were usually quite a few; the 1870 census, for example, showed that of thirteen officers at Fort Laramie, ten were accompanied by their wives.[49]

With wives, came children. At Fort Laramie in 1870, the ten officers' wives had sixteen children, while the seventeen laundresses had thirty-one. In 1880 the 363 officers and soldiers had 70 children of whom 24 were the children of officers. The children rarely had an impact on the larger landscape of the fort, but sometimes devised their own worlds (fig. 3-17). While living in multiple-unit officers' quarters at Fort Laramie, Cynthia Capron mentioned that her children had created a playhouse in the back yard, under the stairs to the upper level.[50]

The number of civilian employees at posts was considerable. The variety of semi-industrial buildings at Fort D. A. Russell in 1870 and 1875 points to the kinds of skilled workmen that were needed: carpenter, butcher, tailor, tinsmith, saddler. Sixty-seven civilian workers and their families lived at Fort Laramie in 1880. In 1870 the household headed by the justice of peace and his wife at Fort Laramie included a brewer, two laborers, a clerk in the quartermaster's department, a telegraph operator, a saddle and hat maker, a wheelwright, a watchmaker, a shoemaker, a freighter, and a butcher. Clearly, all of the trades necessary for a community to function were represented at these posts. Some of these skilled workers had an obvious impact on the landscape, with their workshops clearly delineated on maps of all posts, but others, such as clerks, were less visible. Most of the eleven civilian employees at Fort Bridger in 1866 were teamsters hauling firewood. At Fort Abraham Lincoln, North Dakota, Lt. Col. George Custer condoned construction of buildings for two less necessary but revealing civilian enterprises: a barber shop and photographer's studio.[51]

Fig. 3-17. Fort Laramie, with Laramie River and quartermaster storehouses beyond, 1888. Women and children, shown here on horseback in the Laramie River, were a presence at most posts. Note also the laundry tubs along the riverbank. Courtesy the Brechemin Collection, Fort Laramie National Historic Site.

Every village needs commercial enterprises; army posts differed from real villages in that the army usually permitted only one. The sutler, or post trader as he was known after 1867, operated a commercial establishment much like a general store, providing dry goods, food, tobacco, and liquor to soldiers and officers alike. The secretary of war appointed post traders, allowed them to build their own buildings on the military reservation, and granted them the exclusive privilege of trade. It could be a profitable enterprise; Seth Ward, the sutler at Fort Laramie in 1860, had personal property worth $10,000.[52] With such wealth, the post trader assumed a social position equivalent to the commanding officer.

The power wielded by the post trader could be enormous, as illustrated by William A. Carter, who was traveling with the army when it established Fort Bridger in 1858. He immediately gained the appointment of sutler, although, as a Virginian, he was forced to affirm his loyalty to the Union during the Civil War. In arguing for his job, he mentioned all that he had done in that position: "I have filled the duties of Post Master, Sutler, Probate Judge and U. States Commissioner, since the establishment of this post. I have for the greater portion of the time performed the duties of Pay Master, making out the accounts of the Officers and Soldiers. . . . I have expended all that I have made by close application to my business, in permanent improvements at the Post and in opening [a] large farm and building [a] saw mill for the use of the Government on the Military Reservation." By the 1870s Carter procured goods from every section of the country. Transportation was key; before the transcontinental railroad, Carter ran merchandise wagon trains, bringing his goods overland from a site on the Missouri River. Employing about one hundred men, he received supply contracts for the army, particularly for hay, wood, and livestock. He also supplied shingles, lumber, lime, and lath to the fort's construction projects. When the post was abandoned in 1878, he served as quartermaster agent, and probably lobbied to get the post reactivated (successfully, because the post was reactivated in 1880). His house became a social center, as he cultivated the friendship and goodwill of officers at the post. His son described the scene: "Through his long association with the Army and the marriage of two of his daughters to army officers, Judge Carter's home was looked upon as a center of social life. His excellent library was an attraction and his Steinway square piano that had been hauled across the plains by ox teams, before the building of the railroad, did service not only for dances at this house, but also rendered music from the hands of local artists, as well as distinguished visiting musicians." An 1866 visitor confirmed Carter's social nature, after noting that "he reported his sales at $100,000 per year, and increasing." "He was a shrewd, intelligent man, with a fine library and the best eastern newspapers, who had seen a vast deal of life in many phases on both sides of the continent, and his hospitality was open-handed and generous even for a Virginian." By the time Carter died in 1881, he was one of the most influential men in the western part of Wyoming. After his death, his wife Mary took over his business, operating it until the fort closed in 1890. In the interim she was appointed custodian, then she homesteaded the land and continued to live at the fort.[53]

Fort Laramie did not have a post trader of Carter's longevity or influence, but the post traders there still made an impression on their customers (fig. 3-18). Seth E. Ward was appointed sutler in 1857, and held the post until 1871, most of the time in partnership with William G. Bullock. In 1866 Ward offered to sell his house to the army, which owned the land under it. The sale did not go through, perhaps because Quartermaster General Meigs authorized only $4,000 for the purchase, while Ward demanded $5,000. One visitor described Ward: "We modestly approached the pompous Mr. Ward, who we were told was the sutler. He wore fine clothes, and a soft, easy hat. A huge diamond glittered in his shirt front. He moved quietly round as if he were master of the situation, and with that peculiar air so often affected by men who are financially prosperous and self-satisfied. He seemed to be a good fellow and was in every respect courteous." A young soldier described the scene at the store: "I spent hours in the store

Fig. 3-18. Fort Laramie, post trader's store, ca. 1877. The post trader's store was a hub of activity, attracting Indians, mountain men, emigrants, and settlers, as well as soldiers and officers. Courtesy the Bolln Collection, Fort Laramie National Historic Site.

of the post trader, Colonel Bullock, listening to the conversation and stories told by the mountain men, guides, hunters, and trappers. They all made Colonel Bullock's store their headquarters." He was succeeded by John S. Collins, the post trader during some of the post's busiest years due to the Black Hills gold rush. Collins built a hotel just outside the fort compound, on the stage route. He called it Rustic House, perhaps in an effort to warn visitors that it was constructed of logs, although he promised to "accommodate all with clean beds and first-class meals."[54]

The serving of alcohol was the prerogative of the post trader, although in 1868 the commanding officer at Fort Laramie gave that privilege to W. H. Brown, who ran a "hotel" or mess house across the river from the post, with the following orders: "you are authorized to dispose of to enlisted men a glass of beer three times a day, morning, noon, and evening. You will be responsible that beer is obtained by them at no other time and that the interval between drinks be not less than that described above." Three years later that order was rescinded, and the privilege of dispensing alcohol was granted to post traders. In 1881 President Rutherford B. Hayes banned the sale of alcohol on army posts, depriving the post traders of certain business, and sending soldiers off the reservation to get their drink. Post traders, who continued to sell wine and beer, were supplanted by canteens, which shared their profits cooperatively. Canteens, which were seen as a more wholesome alternative to the post trader's, received official sanction in 1889 to great acclaim.[55]

Post traders sold not only to military personnel, but also to civilians, both those at the fort and those passing by. Particularly at Fort Laramie and Fort Bridger, a significant amount of the trader's sales were made to emigrants along the trail. The post trader served as a magnet, bringing outsiders into the fort, and this constant flow of travelers and traders was an important aspect of the fort. One of the primary groups the trader dealt with were Indians, who passed in and out of the fort along with the rest of the civilians. Margaret Carrington described the vibrant scene at the post trader's store at Fort Laramie during a council in 1866:

> The long counter of Messrs. Bullock and Ward was a scene of seeming confusion not surpassed in any popular, overcrowded store of Omaha itself. Indians, dressed and half dressed and undressed; squaws, dressed to the same degree of completeness as their noble lords; papooses, absolutely nude, slightly not nude, or wrapped in calico, buckskin, or furs, mingled with soldiers of the garrison, teamsters, emigrants, speculators, half-breeds, and interpreters. *Here*, cups of rice, sugar, coffee, or flour were being emptied into the looped-up skirts or blanket of a squaw; and *there*, some tall warrior was grimacing delightfully as he grasped and sucked his long sticks of peppermint candy. Bright shawls, red squaw cloth, brilliant calicoes, and flashing ribbons passed over the same counter with knives and tobacco, brass nails and glass beads, and that endless catalogue of articles which belong to the legitimate border traffic. The room was redolent of cheese and herring, the "*heap of smoke;*" while the *debris* of mounched crackers lying loose under foot furnished both nutriment and employment for little bits of Indians too big to ride on mamma's back, and too little to reach the good things on counter or shelves.

Some Indians worked for the army as guides and were associated with a particular fort or company; others lived nearby and traded at the fort on a regular basis. Brigadier General Cooke ordered post commanders in the Department of the Platte in 1866 to "forbid and prevent the residence in the military reservations of Indian traders, Indian or white. Although Indians are to be treated with hospitality, they must be made to understand that their visits are to be of business and formal. The posts shall not be made centres of demoralization and vice." Nonetheless, Elizabeth Burt noted that Iron Bull, a Crow Indian, was assigned a house inside the stockaded fort of C. F. Smith, Montana, as Iron Bull was the official mail carrier for the garrison in the winter.[56]

Of course, not all Indians had cordial relations with the army. During open hostilities, Indians avoided the forts, preferring to select their sites of engagement. When captured, warriors were often incarcerated at forts. Peace discussions also occurred at the forts, but the numbers participating sometimes rendered the locale impractical. Despite Indians' uneasy relations with the army, they did not always avoid the posts.

Another ingredient in the mix of movement, sounds, and smells at western forts was the presence of animals, providing transportation, food, and companionship. Horses were essential to the cavalry, but large stables and corrals were also devoted to the draft horses of the quartermaster's department. Horses, mules, and oxen pulled freight wagons and had to be accommodated. Cattle provided the 1¼ pounds of beef that was the soldier's daily allotment, although this was often doled out as salted beef, and salt pork was often substituted. Such livestock, horses in particular, were prime objects of theft, as they had to graze far and wide. Warring Indians specialized in running off horses that of necessity strayed beyond the guardians at the forts. Corrals were required to confine horses at night to prevent further theft.

Many descriptions of fort life included dogs, mostly for the fights they engaged in and the noise they caused. Hunting was a common diversion of officers and soldiers, and dogs aided that endeavor. Officers' families brought chickens with them, and often a cow or two for fresh milk. Elizabeth Burt took a cow, rooster, and twelve hens with her from Fort Leavenworth to Fort Bridger to Fort C. F. Smith to Fort Fetterman, giving them up only when her husband was stationed at Fort D. A. Russell, close enough to civilization that she could purchase milk and eggs. Hospitals kept their own set of animals; Fort Laramie's had "two cows, about one hundred chickens, a number of pigeons, and hogs" in 1870.[57]

The sounds of the fort were many. Not just bugle calls, shouted commands, the jingling of spurs and bits, the heavy step of cavalry in motion, but also the Irish accent of a sergeant, the profanity of a laundress, the yelping of a dog, the shout of a child at play, the quiet conversation of an officer's wife making a social call, the notes of a piano from the post trader's house, the well-worn stories of plainsmen gathered outside the post trader's store. Many of these people created the tangible presence of the fort, and all of them acted on and within the landscape of the fort, contributing to its village-like atmosphere.

4

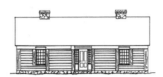

THE HOMELIKE APPEARANCE OF THE HOUSE

Architecture

When Martha Summerhayes accompanied her officer husband to Fort D. A. Russell in 1874, she was appalled by their assigned quarters (fig. 4-1). Each half of the gable-roofed house had three rooms on the first floor, and two rooms in the attic, lit by dormer windows. A porch stretched across the front; a kitchen and servant's room were in a separate building in the rear. When Martha Summerhayes complained that the house was too small, her husband explained that army regulations allocated the number of rooms according to rank, and a lieutenant was allowed only one room and a kitchen; in fact, this house exceeded their allotment.[1]

Martha Summerhayes's second complaint concerned the soundness of the wood-frame house, due to the extremely windy conditions of southeastern Wyoming: "It was early in April, and the snow drifted through the crevices of the old dried-out house, in banks upon our bed." Another army wife, Elizabeth Burt, had had the same experience at Fort D. A. Russell, five years earlier. The dwellings leaked so much that the post commander ordered a board of officers to investigate in 1870. Lt. Col. Thomas Duncan testified that during a storm on March 27 the snow that blew into his quarters stood 5–12 inches deep. Lt. Col. Luther P. Bradley claimed, perhaps hyperbolically, that he removed enough snow from the interior of his quarters to fill three army wagons. And Elizabeth's husband, Capt. Andrew S. Burt, testified on behalf of enlisted men in barracks: "After snow storms at night, the men in the top bunks are covered with snow."[2]

Fig. 4-1. Fort D. A. Russell, double officers' quarters, ca. 1867–1868. Shown here during construction, quarters for cavalry officers were wood frame. Occupants complained about their porousness in the heavy winds of southeastern Wyoming. Courtesy Wyoming State Archives, Department of State Parks and Cultural Resources.

Clearly, the construction of these buildings was inadequate. Officer-turned-novelist Charles King reported that Fort D. A. Russell "had been built at fabulous expense of the cheapest possible materials." The post surgeon described the officers' quarters as being built of "rough one-inch boards placed upright with the cracks battened, the mode adopted for almost all the buildings at the post . . . . The quarters are finished within with planed boards and battens or matched flooring instead of plaster." A lath-and-plaster covering was ruled out, "as buildings here are so much shaken by the wind that plaster is constantly liable to fall off." When apprised of the situation at Fort D. A. Russell, Quartermaster General Montgomery C. Meigs recommended that shingle roofs be replaced with iron, and that tarpaper cover the sheathing boards. This was done; Elizabeth Burt noted that the walls were covered with tarpaper and wallpaper, which "added greatly to our comfort and the homelike appearance of the house." Five years later, though, it still leaked snow during severe storms, to Martha Summerhayes's chagrin.[3]

The shoddy construction and inadequate space associated with these buildings at Fort D. A. Russell reflected the constraints that simultaneously defined fort architecture and contributed to the village-like qualities of the post. Instead of plain, forceful, rectilinear buildings, in the last half of the nineteenth century the fort had buildings that were increasingly picturesque and irregular. And they were still poorly constructed and maintained. But if

> ## Fire at Fort D. A. Russell
>
> Late one night at Fort D. A. Russell, at 5 a.m. on January 5, 1875, with the temperature at 17 degrees below zero, fire broke out in the quarters of Lt. Julius H. Pardee. Fortunately, the blaze took no lives, but the flames destroyed six buildings, or twelve sets of quarters. The board of officers convened to investigate the fire found that all the interior walls of the quarters were of light boards covered with tarpaper, the exterior walls were covered with tarpaper nailed on studding, and the quarters were heated by coal stoves. They concluded that "it would be difficult to devise a more flammable arrangement." Elizabeth Custer, an army wife, had a similar experience with fire and tarpaper. Soon after moving to Fort Abraham Lincoln in Dakota Territory in 1873, she awoke one night to find the house on fire. She sent her eminent husband upstairs to investigate, and thought he was killed when she heard an explosion. As she explained it, "the gas from the petroleum paper put on between the plastering and the outer walls to keep out the cold had exploded. The roof had ignited at once, and was blown off with a noise like the report of an artillery." Shortly after the fire at Fort D. A. Russell, Quartermaster General Meigs prohibited the further use of tarpaper, attributing the ban to the Custers' experience.
>
> Sources: Proceedings of a Board of Officers, 9 January 1875, by order of Brigadier General Ord, RG 92, Entry 225, Box 952, NAB. Custer, *"Boots and Saddles,"* 115. Received from Chief Quartermaster, 20 March 1875, RG 393, Part V, Entry 27, 8, NAB. Drew to Ludington, 25 August 1879, RG 92, Entry 225, Box 533, NAB.

in the earlier period the influences were military ideals and local traditions, in the later period the influences were military tradition and eastern fashions. The plans of officers' quarters tended to be rectilinear, center-passage or side-passage arrangements, as earlier, or sometimes more picturesque cross-wing plans. Beyond the plan, materials and ornament also expressed eastern concepts of civilized living. The preference for wood-frame construction, clad in clapboards or board-and-batten siding, reflects both tradition (the wood frame) and popular modes (the cladding). Similarly, aspects of form and decoration superficially applied to the basic building—such as steep gable or mansard roofs, bay windows, millwork cornices and porch decorations—reflected popular currents back East. Decorative features were more likely to be added in the 1880s, the decade of greatest architectural activity for these three forts. With warfare less likely, duties less onerous, and the surrounding areas more settled, officers adopted more overt expressions of eastern civilization. The collection of buildings showed a wide variety of style and materials. While a village reflected planning, in maintaining an open space at the center, private construction around that open space resulted in unplanned variety. Although forts operated under different conditions than private development, in some cases the result was the same.

Like others across the West, by 1890 these three forts were disorderly assemblages of buildings. Their original layouts had been stretched and disrupted. Their buildings reflected different construction programs and different ideas about appropriate military buildings. The factors involved included regulations regarding space, which fostered innovative solutions; improved transportation systems, which expanded the range of available building materials; and a somewhat stubborn rejection of professional expertise and centralization, which meant that each post retained an individual appearance. At the same time, a creeping tendency toward modernization resulted in some surprising changes: regulations eventually mandated contracting out construction projects, so that professionals did the work; technological experimentation increased the range of available building materials; and standardized plans for hospitals and nonresidential buildings were widely adopted. The village quality of the architecture is due mostly to this mixture of new and old, well built and shoddily constructed, brick and wood frame.

As Built

An examination of the buildings constructed in this period again reveals rational plans but uneven execution. Fort D. A. Russell began as a post with an exceedingly regular appearance. Set in an unusual diamond-shaped plan, the board-and-batten buildings on the parade formed a uniform, military environment. Seven double officers' quarters on two sides of the parade balanced six barracks on the opposite two sides. The office of the chief quartermaster of the Department of the Platte probably produced the designs, just at the time it was furnishing plans for use department-wide; the Fort D. A. Russell buildings may represent those plans. All of them were wood frame with board-and-batten siding, probably built by the troops themselves. Because they were double houses, the officers' quarters were rigidly symmetrical (fig. 4-2). Each half had an entrance in the end bay, which led to the stair and hall; three rooms made up the first floor. While the second floor had two large rooms and two closets, the slope of the gable roof reduced usable space.

The 80-by-30-foot barracks had a large, open dormitory intended for eighty soldiers sleeping in double bunks, heated by three stoves. Although the wood frame of the buildings had been infilled with adobe bricks, the building was still porous to wind and snow. Behind each barracks was a

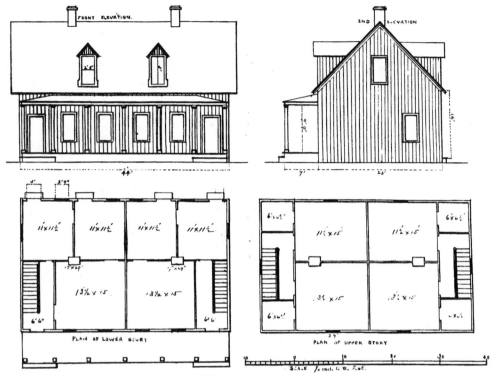

Fig. 4-2. Fort D. A. Russell, double officers' quarters, plans and elevations, ca. 1867. Each officer at Fort D. A. Russell was allocated three rooms, not counting the rooms in the attic or the kitchens that were soon joined to the rear. The front room served as the parlor, the dining room was behind it, and the room behind the stairs served as the kitchen or, later, access to the kitchen wing. Bedrooms were in the attic. From Medical History of the Post [Fort D. A. Russell].

mess hall and kitchen building constructed of "logs placed upright in the earth."[4]

Set in front of the guard house, near the bottom point of the diamond-shaped parade ground, was an unusual and picturesque structure—a hexagonal, three-tiered guard tower (fig. 4-3). The board-and-batten-sided structure had a porch around the first floor, and the second and third stories were progressively smaller. Crenellations and a stovepipe crowned the top. The first story provided a room for the officer of the guard, and the third story, reached by an exterior stairway, served as a lookout for the sentinel. Behind this eccentric guard tower was a more ordinary guard house for prisoners. The 40-foot-square building was constructed of a wood frame sided with boards and battens, like the other buildings at the post.[5]

The 1875 site plan showed several buildings off the parade ground, including shops, stables, laundresses' quarters, and special-purpose buildings. Little documentation of these buildings survives. One mention in the

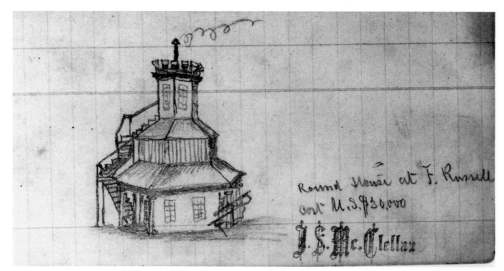

Fig. 4-3. Fort D. A. Russell, sketch of guard tower, ca. 1873–1877. The unusual hexagonal guard tower is pictured with smoke rising from the chimney in the center of the crenellated roof. From McClellan, James S., U.S. Army, 3rd Cavalry, Diaries, 1873–1877, Manuscripts and Archives Division, The New York Public Library, Astor, Lenox and Tilden Foundations.

surgeon general's report of 1875 implied that they were minimal dwellings: "The quarters for married soldiers and laundresses now number forty-six sets, scattered to the south and west of the post. The greater part are built of slabs, placed on end, and consist each of two rooms and a kitchen. Ten of the number are built of adobes, but of the same size and plan as the wooden buildings."[6] For the most part, the army devoted its attention to the more important buildings on the parade ground.

The symmetry of the buildings on the parade ground was first interrupted in 1875 by the fire that destroyed six double officers' quarters, leaving the northwest side of the parade ground vacant. That year, only two companies and eleven officers occupied the post, so there was no immediate housing shortage. In 1882 General Sherman put Fort D. A. Russell on the list of permanent forts, and the next year the post housed seven companies, a level it sustained or exceeded until the Spanish-American War. With the complement of men and officers staying above three hundred, and the number of companies ranging from seven to nine, the post required expansion. And with Sherman's policy of establishing permanent posts, Fort D. A. Russell's new buildings merited brick construction.

The 1885 building program included six company barracks, six captains' quarters, and six quarters for noncommissioned officers, all of brick. The barracks were one story with gable roofs and porches across the front (fig. 4-4). Measuring 104 by 30 feet, they were divided into a 93-foot-long dormitory and two small rooms for the sergeant and storage (fig. 4-5). Rear

Fig. 4-4. Fort D. A. Russell, brick barracks, ca. 1895. Six brick barracks were built on the parade ground in 1885 to replace frame barracks, which were moved and used as rear wings. Courtesy National Archives.

additions, forming a T-shaped or L-shaped plan, consisted of frame barracks or storehouses that had been built in 1867, moved, and adjusted to create space for a dining room, kitchen, cook's room, pantry, and wash- and bathrooms.[7] In 1896 these frame wings were replaced with brick ones that formed U-shaped plans (fig. 4-6). The new wings contained a day room, mess room, and kitchen on one side, and additional dormitory space and a complete bathroom on the other.

Across the parade ground, six captains' quarters went up in 1885, followed by four more two years later. The designs for these apparently originated at department headquarters in Omaha, Nebraska, as very similar plans were drawn there for Fort Omaha in 1882. Capt. James Lord, assistant quartermaster, supervised construction of the first four, while Capt. Charles F. Humphrey, depot quartermaster, was called in to oversee the next four.[8] Jigsawn woodwork decorated the porches across the fronts of these picturesque dwellings, whose massing was broken by wall dormers (fig. 4-7). Described as being one-and-three-quarters stories, the captains' quarters had a side-hall entrance and three rooms on the first floor of the main block, with a kitchen in a wing behind (fig. 4-8). This exceeded the allotment for captains, but there was even more space on the second floor: two bedrooms,

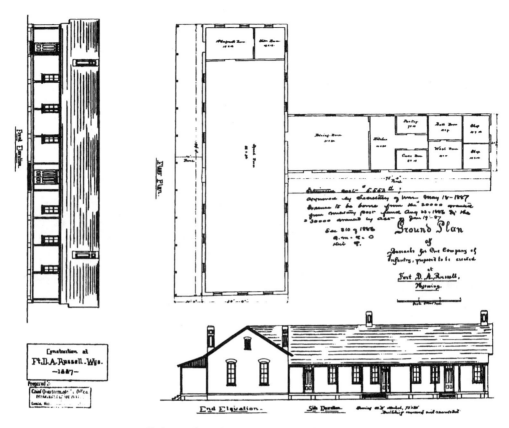

Fig. 4-5. Fort D. A. Russell, barracks, plan, 1887. An additional brick barracks constructed in 1887 also reused a frame building for the rear wing. Courtesy National Archives.

a servants' room, a child's room over the front stairhall, and a storage loft over the kitchen. The second floor, although consisting of a half-story, had a floor-to-ceiling height of 7½ feet at the exterior walls, rising to 10 feet just 3½ feet from the exterior.

In plan, the 1885 captains' quarters were similar to the half of a double officers' quarters that preceded them. Each had a side-hall entrance, with a wider room behind, and a wide room in the front, with a narrower room behind. Kitchen and pantry were added onto the earlier construction, but incorporated into the later structures. The side-passage plan, which Henry Glassie termed "two-thirds Georgian," retained the basic design framework of the Georgian plan.[9] The 1885 captains' quarters benefited from increased room size, height of the second floor, and brick walls compared to board and batten; in addition, they were single, not half of a larger structure. All this made them far grander and more comfortable than their predecessors.

The room sizes of the brick captains' quarters compared favorably to

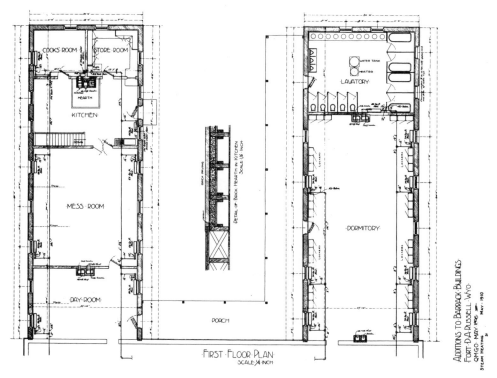

Fig. 4-6. Fort D. A. Russell, barracks, plan of rear wings, 1896. A decade later the frame rear wings were replaced with two brick wings, forming a U-shaped plan and providing fully equipped bathrooms. Courtesy F. E. Warren Air Force Base.

civilian buildings as well. A group of nine brick houses built in the 1890s in downtown Cheyenne provides an interesting contrast. Designed by architect J. P. Julien and built by Moses P. Keefe, a contractor who built much of Fort D. A. Russell, the civilian buildings had the same general plan as the captains' quarters, although with the advantage of a bathroom. Far more elaborate on the exterior, the civilian buildings were much smaller within: the main rooms of the captains' quarters were more than 40 percent larger than those in the house at 414 E. 22nd St. (fig. 4-9). The unified designs of the civilian buildings recalled the fort buildings, and they probably appealed to the middle-class professional equivalents of captains, but their fancy exteriors were not matched by spacious interiors. The army apparently was more willing to spend money on large rooms than on fashionable decoration.

The original wood-frame officers' quarters, now twenty years old, still faced the parade ground. In 1883 one side of each of these buildings was weatherboarded, and each set of quarters received a storm shed—a vestibule on the exterior of the house, under the porch. In 1885 they received new

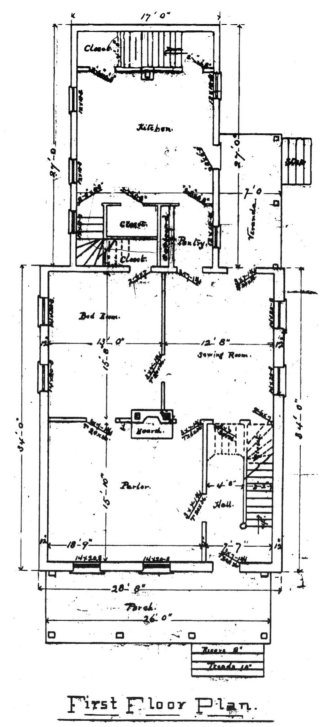

Fig. 4-7. Fort D. A. Russell, captains' quarters, first-floor plan, 1887. The four brick captains' quarters built in 1888, identical to those built three years earlier, had a floor plan similar to the original quarters (fig. 4-2), but set in single dwellings and with much greater dimensions. Courtesy National Archives.

Fig. 4-8. Fort D. A. Russell, captains' quarters, ca. 1890. Constructed in the 1880s, the row of eight brick officers' quarters stretched eastward off the original diamond-shaped parade ground. The wall dormers yielded a generous 10-foot ceiling height on the second floor. Courtesy Wyoming State Archives, Department of State Parks and Cultural Resources.

floors, painting inside and out, and wallpaper. In a further attempt at creating uniformity, the various back buildings and sheds were torn down. Charles King described the building that had gone on in the back yards of these quarters: "The kitchen was in an annex to the cottage. The servants' room in an annex to the kitchen. The coal-shed, wood-shed, the trunk- and box-sheds were annexes to the servants' quarters, and the cow-shed was an annex attached impartially to the rearmost end of the trunk-room and to the inner side of the high fence of rough, unpainted, weather-warped boarding that separated the premises from the bleak wastes of prairie that stretched far away northward." King blamed parsimonious appropriations for the makeshift construction: "Man after man the successive post quartermasters had yielded to the importunities of the occupants and furnished plank and scantling, spikes and nails; and year after year had this pennywise patchwork been going on, until the result was a veritable architectural crazy quilt minus all that makes this quilt attractive,—its bright variety of color. Nothing on earth was ever much more unpicturesque than a rear view

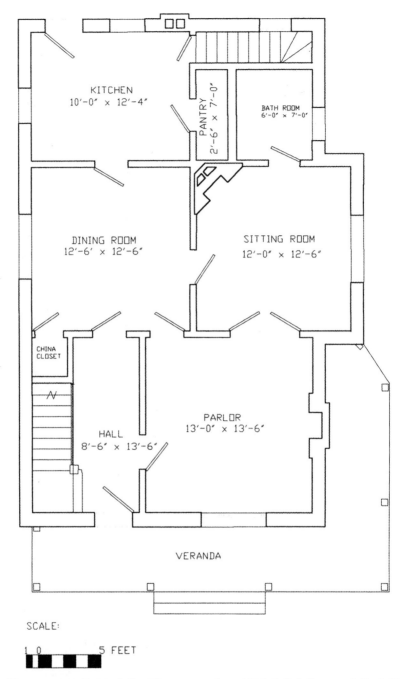

Fig. 4-9. House at 414 E. 22nd St., Cheyenne, plan, 1892; J. P. Julien, architect. Equipped similarly to the captains' quarters and built just a few years later, a group of brick houses in Cheyenne offers a downstairs bathroom but much less space overall. Drawing in possession of Ed Tarbell, redrawn by author.

of the fort." These various back buildings were replaced with wings, 12 feet wide and 43 feet long, containing a dining room, kitchen, and storeroom. The post quartermaster called them "a very great improvement, giving the buildings a uniform appearance throughout, and helping greatly sanitary conditions." In 1888 these older officers' quarters were placed on stone foundations. With these renovations, they lasted until the turn of the century.[10]

Despite the new construction in the 1880s, many of the original wooden buildings remained, both in service as intended and in new functions, as back wings to new buildings. Fort D. A. Russell persisted in being an irregular, unstandardized post, despite its orderly original design. It was not until the next building program, in the early twentieth century, that all of the old wooden buildings were demolished and the post began taking on the cohesive brick-built appearance that it has today. The diamond-shaped parade ground began its spread toward the east in the 1880s, evolving into an elongated rectilinear parade ground with irregular points and circles protruding from it.

Forts Laramie and Bridger also continued to evolve in this period. At Fort Bridger, a few buildings were added in the 1870s, but the general appearance of the post did not change. The army temporarily abandoned Fort Bridger in 1878, regarrisoning it in 1880 with two companies, totaling seventy-six men and officers. Although in 1882 General Sherman put Fort Bridger on his list of "temporary" posts, with a projected life of ten years, the fort received new attention due to the activity of the Utes in western Wyoming. The army initially wanted to enlarge Fort Thornburgh, southwest of Fort Bridger in Utah, but could not acquire the land, so it abandoned Thornburgh and expanded Fort Bridger to a five-company post.[11]

With $33,500 in construction money allocated in 1883, 2nd Lt. Charles P. Stivers, post quartermaster, oversaw the construction of ten buildings, including a set of barracks, commanding officer's quarters, and two sets of officers' quarters, all wood frame.[12] The commanding officer's quarters was particularly grand: a large two-story house with a one-and-a-half-story wing set on stone and concrete foundations. The porch across the front, bracketed eaves, hoodmolds over the windows, and bay windows were in stark contrast to the low, one-story log buildings of twenty-five years earlier. But the commanding officer's house had an interesting parallel at the fort—the post trader's house.

Post trader William Carter had held his position since the establishment of the fort, and he wielded considerable influence in the western part of Wyoming. His hospitality was legendary, as attested by Elizabeth Burt, who arrived with her husband, appointed commanding officer, in 1866:

We drove to the quarters of the commanding officer, lately vacated by a volunteer and his family, who were still in the post. An invitation awaited us to dine with Judge and Mrs. Carter. This was gladly accepted and highly appreciated. An excellent dinner in a most comfortable dining room showed how attractive life might be in a one-story log house, even in far away Utah, the home of the Mormons.

The Judge was the post trader and a most agreeable gentleman, while his wife and family charmed us by their kind welcome and gracious manners, creating the feeling that we had found congenial companions who would prove good friends.

It was no accident that Judge Carter set himself up as an equal to the commanding officer, whose wife described Carter as a "gentleman" and his family as potential "friends." Carter's wealth far exceeded that of any army officer, but his position depended on maintaining good relations with the army. His house became an expression of his position, not only for the hospitality that went on inside, but also for the statement made by its architecture. A few years before officers' quarters at the post gained fashionable millwork ornamentation, the post trader's house acquired decorative hoodmolds over the windows, two bay windows, and mill-sawn siding, although the house remained a low one-story log structure (fig. 4-10). After Carter died in November 1881, his wife Mary carried on the family business. In the 1880s the Carters' house and the commanding officer's displayed a similar fashion sense.[13]

New building continued at the post. Two stone warehouses constructed in the 1883–1884 boom were joined four years later by three more stone buildings: barracks, guard house, and oil house. A fireproof building for the storage of "mineral oil," or kerosene, was deemed essential. The oil house was a pyramidal-roofed building 20 feet square, located near the new stone storehouses northeast of the parade ground. On the west side of the parade ground, where storehouses had formerly stood, was the site of the new stone barracks, T-shaped in plan (fig. 4-11). Contractor George H. Jerrett of Fort Duchesne, Utah, built these three stone buildings for $8,500.[14]

Despite this effort at permanent construction, Fort Bridger remained on the "to be abandoned" list, and the army withdrew its forces in 1890. At that time, an estimate of the value of the buildings at the post listed twenty-one buildings of log, fourteen of frame, and eight of stone, ranging in value from $10 (a stone magazine) to $4,000 (stone barracks).[15] Officers' quarters ranged in value from $3,500 for a new double house to $150 for a three-unit, log-and-frame building. Nonmilitary housing included two buildings for laundresses and one for teamsters. Only three of the buildings at the

Fig. 4-10. Fort Bridger, post trader's house, ca. 1910. The low log house, which served as a social center of the post, was updated with bay windows and hoodmolds. Courtesy Wyoming State Archives, Department of State Parks and Cultural Resources.

post were two-storied. The architectural cohesion of the early years had been lost in the updating and modernizing of the 1880s.

Fort Laramie underwent similar changes. In 1874 the *Omaha Herald* described Fort Laramie as "very peculiarly built": "There are wooden houses, and adobe houses, and concrete houses. There are new, pretentious looking houses, and good, substantial looking houses, and tumble-down, dilapidated looking houses. . . . I suppose the greatest combination of the grotesque, the filthy, the ill-arranged in the department is here at Ft. Laramie."[16] Into the 1870s Fort Laramie remained close to the center of the conflict with the Indians, but new construction at the fort gradually produced sounder and neater residences. By the 1880s, as the number of enlisted men at the fort dropped below two hundred, Fort Laramie became a more tranquil post. Trees and street lamps added to the domestic feel, and new concrete buildings tended to modernize the appearance of the place. By the time the army abandoned Fort Laramie in 1890, the post had a variety of buildings, constructed over a forty-year period. Building materials

Fig. 4-11. Fort Bridger, stone barracks, 1924. The stone barracks built in 1887–1888 saw only a few years' use before the post was abandoned in 1890. Courtesy Wyoming State Archives, Department of State Parks and Cultural Resources.

ranged from wood frame to adobe to concrete. Some buildings were two-story, upright structures, while others were low, one-story "huts." Some were new and sound, while others showed the effects of age.

Space

Army regulations regarding the number of rooms allotted to each officer by rank remained unchanged in this period, but ways of manipulating space multiplied. There was an architectural solution to the problem of how to increase the number of rooms in a dwelling without appearing to do so: the inclusion of habitable attic spaces. Army regulations specified that attics did not count as rooms, so creative post quartermasters designed quarters with livable attics. Model plans issued by the Quartermaster General's Office in 1872 made use of wall dormers and windows on the gable ends to light attic spaces. Post quartermasters soon realized that the mansard roof—a style that was sweeping the country in the post–Civil War period, coinciding with extensive military construction on the Great Plains—was an even better solution, providing adequate head room for most of the second floor (fig. 4-12). Not all observers liked the mansard roof, however. Lt. Col. Thomas M. Anderson condemned it: "Why Mansard roofs? Why transfer this unsightly Paris abomination to the wilderness? All good architects

Fig. 4-12. Fort Laramie, officers' quarters, section, drawing by Leslie E. Wilkie, 1939. The mansard roof created the most space on the second floor while still enabling it to be counted as an "attic" and therefore outside the room allotments. In this quarters at Fort Laramie, the mansard roof produced a generous floor-to-ceiling height of 9 feet 10 inches. Courtesy Library of Congress, HABS.

condemn them. The fire commissioners of every civilized city have protested against them as fire-traps." Nonetheless, mansard roofs and a variety of dormers decorated the quarters at Fort Laramie. Anderson called the attic rooms "a kind of official evasion" of the room allocation regulations. Another officer confirmed this tacit understanding: "By common consent the half story above the ground floor is by a stretch of commanding officer–quartermaster–inspector conscience counted an attic."[17]

Another way to increase allotted space was to enlarge rooms. In 1867 2nd Lt. Frank A. Page pointed out to his commanding officer that "officers' quarters are frequently built so small as to allow a Lieutenant but very little more space than is allowed to enlisted men by Regulation." The Marcy Board, a panel convened in 1871–1872 to revise army regulations, suggested that no room in officers' quarters be less than 15 feet wide or deep, but this regulation was never adopted officially.[18] Still, rooms in officers' quarters seemed to increase in size. The first-floor rooms of the double officers' quarters constructed in 1867 at Fort D. A. Russell were 15 by 13½ feet, 11 by 11½ feet, and 11 by 11½ feet. First-floor rooms of the single brick quarters that replaced them in 1885 were 17½ by 15½ feet, 12½ by 16½ feet, and 13 by 16½ feet, or about half again as large.

Nonetheless, officers continued to object to their allocation of rooms. Even Quartermaster General Meigs objected that the allocated rooms were too few. After the fire that destroyed twelve sets of officers' quarters at Fort D. A. Russell in 1875, Meigs submitted a design for replacement buildings. Taking this opportunity to criticize the room allotments, he claimed that if he gave officers space according to regulation, "This would reduce the quarters for a lieutenant to those which are generally asked by commanders

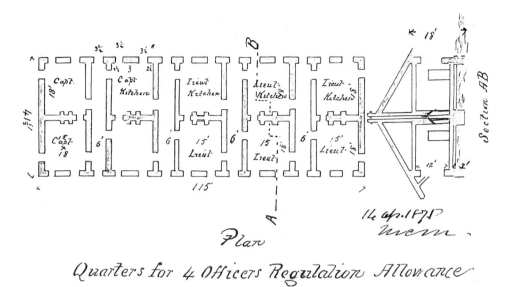

Fig. 4-13. "Quarters for 4 Officers Regulation Allowance," plan, 1875. When Quartermaster Gen. Montgomery Meigs was forced to obey the room regulations, he produced this plan for a four-unit building. Three lieutenants received their allotted one room and a kitchen, while the captain claimed two rooms and a kitchen, leaving one room unassigned. Courtesy National Archives.

of posts and depots for a laundress, viz: two rooms on the ground floor without attic or cellar or kitchen." The regulations, he felt, allowed too few rooms: "But to require officers who are gentlemen, and who have respectable families to live in two small rooms during the whole period of their active lives until by promotion to rank of captain they become entitled to three such rooms, I think has never been the custom of the service." He further protested: "To enforce such a rule will be the cause of the greatest discontent and suffering. It will in fact separate officers at the frontier posts from their families, who cannot live decently thus mingled, old and young, male and female, children, servants, masters and mistresses, in two rooms." When Inspector General Randolph B. Marcy refused to be swayed, Meigs grudgingly produced a plan according to the regulations. The plans provided for a 115-by-41-foot, one-story building of twelve rooms, arrayed two rooms deep with 6-foot-wide transverse halls (fig. 4-13). Marcy claimed the last word, asserting that he had served twenty-seven years as a line officer in small quarters, and he had never complained.[19]

As Meigs hinted, the room allocations for officers were routinely ignored, not only through devious means, such as manipulation of attic spaces and room sizes, but also through outright construction of quarters that provided more rooms than were allowed. A few months before he was appointed quartermaster general in 1883, Samuel B. Holabird tackled this problem in the pages of the *Army and Navy Journal*. In arguing for the presence of ladies at a post, he recognized that more rooms should be allocated. "The lowest officer in grade in service needs, for the health and comfort of himself and family, to have two main rooms en suite, besides his kitchen. . . . Over these rooms ought to be two chambers with a suitable staircase arranged to reach them, and in the basement sufficient storage room, etc." Holabird also came out against attic rooms, arguing instead that "the rooms that count in the assignments should be those only upon the main floors; dormitories in the second story and basement rooms, as such, should not be counted."[20] Once Holabird was named quartermaster general, two-story officers' quarters became the norm, and the pretense of attic rooms was dropped.

Noncommissioned officers (NCOs) traditionally lived in barracks. The first sergeant might receive a separate room, but the others slept in the dormitory. In the 1870s and 1880s several post NCOs were added to the usual complement of company NCOs and regimental NCOs. Hospital stewards, ordnance sergeants, and commissary sergeants were assigned to most posts by 1881, as were post quartermaster sergeants in 1884. This growth in the NCO ranks created a middle class that, although small, was highly visible on the landscape. Without a company assignment, these NCOs required separate housing. In 1885, the army constructed six identical quarters at Fort D. A. Russell for noncommissioned staff officers (fig. 4-14). The one-story, three-room brick buildings measured 22 by 27 feet. The floor plans were changed over time, but apparently had an arrangement similar to one unit of the double officers' quarters built in 1867. A year after construction, the post quartermaster described the NCO quarters as inadequate: "They are small and comfortable for man and wife, but where there are a number of children in the family they are too small and decidedly uncomfortable and unhealthy, but this defect in case of a large family can be remedied by a small extension." He requested authorization for just such an addition for a commissary sergeant with four children, but it was disapproved without comment.[21]

Also in 1885, Fort Laramie had tried a different approach to housing its noncommissioned officers: one long, one-story concrete building. Measuring 242 by 27 feet, the quarters was divided into six four-room units, each with a central door. Although not as domestic in appearance as those at Fort D. A. Russell, the new quarters did provide considerably more space. These quarters were assigned to the ordnance sergeant, commissary sergeant, regimental quartermaster sergeant, principal musician, post quartermaster sergeant, and chief musician.[22]

138 ■ VILLAGE: FORTS LARAMIE, BRIDGER, AND D. A. RUSSELL, 1867–1890

Fig. 4-14. Fort D. A. Russell, NCO quarters, ca. 1895. Constructed in 1885, brick cottages for noncommissioned officers assigned to the post had jerkinhead gable roofs and porches across the front. Courtesy National Archives.

Personnel

While regulations allocating space remained fixed, encouraging subterfuge, regulations regarding construction changed fundamentally in a move toward permanence and modernization. Recognizing that his troops were inadequately sheltered, General Sherman in 1882 proposed a radical reorganization of the army in the West, concentrating his troops at fewer posts and abandoning the rest. Although there were compelling economic and tactical reasons for this new alignment, improved quarters were also one of Sherman's goals. In his annual reports of 1880 and 1881, Sherman deplored the fact that "permanent buildings" required the approval of Congress. "The time for temporary shanties has passed away, and no building should henceforth be erected at any of the permanent forts and military establishments except of stone or brick." In 1882 he put forth his lists of permanent posts, temporary posts with a projected ten-year life, and posts to be abandoned, proposing that the permanent forts would have "barracks of stone or brick . . . so as to quarter troops, which will otherwise soon be roofless by the decay of the temporary quarters which have heretofore been their homes."[23]

Quartermaster General Samuel B. Holabird oversaw the implementation of this new plan of concentration. In his articles in the *Army and Navy Journal*, Holabird outlined his philosophy. He particularly advocated spacious and handsome officers' quarters in order to provide for officers' families who, he felt, served as a civilizing influence both for the men and for the population in the surrounding area. He insisted that "all the structures should be of stone or brick, and all roofing materials of metal." He rejected the idea of standard plans, noting that "it is found in practice extremely difficult to frame any one plan to suit all possible climates and situations. In fact, it is practically impossible. Each plan *must grow out of the necessities of the case*, and thus be perfectly adapted to its surroundings."[24]

To improve construction practices, Holabird advocated contracting the work out to civilians and using a construction supervisor from the quartermaster's department. As he explained it, "The best results appear to have followed where there has been an officer assigned to the work, and where the work has been accomplished as far as practicable by means of the contract system; especially to secure the materials. Of course there are some individuals competent and successful as superintendents, but there are others less so, and on the whole the contract system as provided by law is by far the most economical." Congress responded to Holabird's suggestion. The House added the following language to the army appropriations bill for fiscal year 1885, debated in the spring of 1884: "The erection, construction, and repairs of all buildings and other public structures in the Quartermaster's Department shall, so far as may be practicable, be made by contract after due legal advertisement." The Senate Appropriations Committee recommended that this provision be deleted, but it was reinstated during reconciliation, resulting in far-reaching consequences for how army buildings were designed and built.[25]

Even before this dramatic change, chief quartermasters from the department had issued plans, and assistant quartermasters, also from the department, had been assigned to posts to oversee construction programs. Now, the assistant quartermasters or post quartermasters contracted construction to civilian firms, a practice that led eventually to even more centralized designs.

Building Materials

The railroad, the great agent of change in the West, transformed military supply systems and made importation of building materials more feasible.

Alien materials, and sometimes whole buildings, were imported for forts, and when this was impracticable, technological solutions were explored. The simplest process was to bring wood to treeless areas. Fort D. A. Russell, with no timber of its own to draw on, nevertheless sported wood-frame buildings. The army contracted to buy lumber, and the fort's location on the Union Pacific rail line solved the transportation problem.[26] When the army shifted to brick construction in the 1880s, it looked to communities both near and far to augment its own brick production.

The decision to rebuild Fort D. A. Russell in brick in the 1880s was a momentous one. In his recommendations made in 1882, General Sherman noted that new construction at the permanent forts should be brick or stone "of the most permanent character, meant to last forever." This was echoed the following year by Brig. Gen. Oliver O. Howard, commander of the Department of the Platte, who recommended that Fort D. A. Russell "be completely rebuilt of brick or stone, having buildings as good, for they should be as permanent as those for the United States courts or the United States post office."[27] Over the next three decades, Fort D. A. Russell underwent the transformation that Howard suggested.

Like stone, brick was a permanent, fire-resistant material, and it could sometimes be produced on site. Soldiers often manufactured bricks for chimneys, but bricks had to be fired in a kiln, and the fuel required was sometimes difficult to obtain. As Lieutenant Woodbury reported about Fort Laramie in 1851, "Bricks burnt near the site, but wood must be hauled five or six miles unless drift accumulates at nearer points." In addition, the labor required to produce the quantity of bricks for a building was too great to make brick a popular choice at temporary posts. When Fort Totten, North Dakota, was built in 1869, it took an act of Congress to permit the buildings to be constructed of brick.[28]

Once Fort D. A. Russell was determined to be a permanent post, however, brick building proceeded apace. By this time Cheyenne brickyards were producing bricks for local consumption; the newspaper reported in 1882 that "the two brick yards in this city have made 3,000,000 brick and early in the season the scarcity of this material led to its importation from Colorado points." Nearly all of the buildings constructed at Fort D. A. Russell in the late 1880s and after were of brick, mostly obtained locally. The vast scale of the military construction program, producing twenty-one new buildings in 1885, was apparently more than local businesses could supply. In 1888 the assistant quartermaster noted that "the supply of brick on hand in this vicinity is about exhausted, and until new kilns are prepared and burnt it will be impracticable to proceed with the work."[29] Whether taking advantage of its neighboring city or shipping bricks in from Colorado, Fort D. A. Russell entered a modern era when it shifted to more permanent brick construction.

The presence of the railroad gave rise to the notion that prefabricated buildings might be a solution to the absence of local building materials. In one sense, the army had always carried its own prefabricated shelter—the tent. The canvas-covered, elongated triangular tent was supplanted by the conical Sibley tent, introduced in 1858; Quartermaster General Thomas Jesup called it "a copy of the Sioux lodge," or tipi.[30] With the addition of vertical walls and board floors, the tent served as soldiers' and officers' primary dwellings.

The army was less successful with more permanent, prefabricated construction. Capt. Parmenas T. Turnley was one of the first to develop a prototype, which was of plank construction. Assemblers inserted vertical pine boards into a grooved sill and capped them with a grooved plate. Smaller battens between the boards were grooved on both sides, holding the planks in place. A shallow-pitched gable roof and prebuilt windows and doors completed the structure. In 1855, thirty-seven of these "Turnley cottages" arrived by steamboat at Fort Pierre, South Dakota. Barracks, officers' quarters, hospital, and storehouses all made use of the same construction system. Unfortunately, Fort Pierre endured one of the coldest winters on record, exposing the "frail" quality of these drafty quarters. Even portable cottages sent to Camp Lancaster, Texas, were deemed too cold and drafty. The cottages could not be used without extensive reinforcement and insulation, and Turnley's experiment was unsuccessful.[31]

In 1867 William Myers, chief quartermaster of the Department of the Platte, suggested portable officers' quarters and storehouses. His supervisor, James L. Donaldson, the chief quartermaster at the Division of the Missouri, forwarded the suggestion to Washington, pointing out that when materials come from great distances, transporting buildings might not cost much more: "The cost of transporting them when taken apart would be very little if any greater than the transportation of the lumber required in their construction." Myers also received an inquiry from his colleague, the chief quartermaster at the Department of the Missouri, asking for a plan of the storehouse and confessing that he had tried "to get up something of the kind but can get nothing that is portable that will stand the high winds."[32]

Whether prefabricated buildings were shipped to Fort D. A. Russell is not known. Myers's correspondence concerning "portable" buildings coincided with the development of Fort D. A. Russell in the summer of 1867. Elizabeth Burt, who accompanied her husband there in 1869, remembered the officers' quarters as having been fabricated in Chicago and shipped to

Cheyenne, but she wrote many years after the fact and was the only one to identify them as such. In a report in 1870 post surgeon C. H. Alden, who had been at the fort during construction, identified only the commissary and quartermaster's storehouses as prefabricated. Long wooden buildings, 25 by 100 feet, the warehouses were in "the style known as sectional, they having been brought up from Omaha in parts, and put together here."[33]

These may have been the storehouses that Myers developed, but shipment of prefabricated buildings from one construction town to another along the railroad line was not unusual; many of Cheyenne's early buildings had been shipped from Julesburg, the previous terminus of the line.[34] Shipment to forts off the railroad line was another matter, despite Donaldson's claims to the contrary. Fort Laramie was located 89 miles from the railroad, so freighting costs would outweigh any gains in labor saved. Even at Fort Bridger, only 10 miles from the railroad, prefabricated buildings were not employed.

Fort Laramie did receive one dramatic example of a manufactured structure: an iron bridge across the North Platte River (fig. 4-15). In 1873, in the midst of a growing interest in gold in the Black Hills, Cheyenne citizens became concerned that freighting traffic might leave the railroad at Sidney, Nebraska, to cross the North Platte north of there, rather than depart from Cheyenne and take the chancier crossing at Fort Laramie, which was unfordable two or three months of the year. At the behest of his constituents, territorial delegate W. R. Steele introduced legislation in Congress that appropriated $15,000 to build a bridge at the Fort Laramie crossing. When that legislation passed in June 1874, the Department of the Platte awarded the contract to the King Bridge & Manufacturing Company of Cleveland, Ohio, the only bidder to propose iron. The wrought iron bridge, which reached 420 feet in three spans, traveled in pieces by rail to Cheyenne, then overland to Fort Laramie. Capt. William S. Stanton of the Engineer Department supervised the assembly and tested the bridge when

Mixing Concrete

In 1873 the Quartermaster General's Office published *Notes on Building in Concrete and Pisé*, which contained the following guidelines for constructing "grout" walls:

> Grout (or concrete) as building material, is composed of lime and coarse gravel, mixed with water, in the same manner as ordinary mortar, and of about the same consistency.

> A small proportion of cement—say, one-tenth—may be mixed with lime, and will add to the strength, durability, and finish of the walls.
>
> To give solidity, quarry chips can be used by being imbedded in the composition while forming the walls.
>
> The walls of buildings are formed of this material by making boxes or molds of 2-inch plan, of uniform width, placed on edge horizontally, as wide apart as the desired thickness of the walls, and held in place by cleats. Into these boxes or molds thus formed the grout is put, and allowed to remain a sufficient time to harden; when the boxes are removed, and again placed at the top of the layer just formed, and again filled with gravel and stone and so on until the wall is of the required height. . . .
>
> The gravel may be mixed with a moderate proportion of sand advantageously.
>
> Lt. W. W. Rogers, posted to Sidney Barracks, Nebraska, supplied this information to the quartermaster general, but the publication omitted the crucial information of the proportion of lime to gravel. At Sidney Barracks, Rogers had first used one part lime to twelve parts coarse gravel and sand. In subsequent concrete buildings he added cement to the mixture and recommended one-fifth part. There were varying accounts of the formula that Fort Laramie used: one part lime to twelve parts sand and one-half part cement, or one part lime to ten parts sand and one-quarter part cement. At any rate, it was successful; most of the concrete walls poured in the 1870s still stand today.
>
> Sources: *Notes on Building in Concrete and Pisé for the Frontier.* Rogers to Perry, 23 January 1873, Fort Laramie vertical file HB-5, Fort Laramie National Historic Site. William P. Hall to Secretary of War, received at Military Division of Missouri 5 August 1880, RG 92, Entry 225, Box 532, NAB. Anderson, "Army Posts, Barracks and Quarters," 441.

it was completed, by loading thirteen army wagons with stone and leaving them on the bridge for several days. Pronouncing the bridge safe and secure, he opened the bridge to the public, and it still stands today.[35]

In an effort to find a suitable building material for treeless, railroadless posts like Fort Laramie, the quartermaster general issued a pamphlet in 1873 touting two materials: pisé, or rammed earth, and concrete. The quartermaster general's explanation of concrete derived from the experience of 1st Lt. William W. Rogers, acting assistant quartermaster at Sidney Barracks, Nebraska, in 1872. Rogers explained that a mixture of lime and coarse gravel combined with water was poured into forms to create walls. Rogers also noted that "a small proportion of cement—say, one-tenth—

Fig. 4-15. Fort Laramie, iron bridge, 1996. Fabricated in Cleveland and assembled near Fort Laramie in 1875, the bridge enabled gold prospectors to cross the North Platte River on their way to the Black Hills. Author photo.

may be mixed with lime, and will add to the strength, durability, and finish of the walls." Rogers credited the superintendence of "Mr. L. Hobbs, master mechanic," in the construction of the laundresses' quarters at Sidney Barracks, but it is unclear whether the introduction of concrete construction was due to Rogers or Hobbs. There is one earlier reference to a concrete building, at Fort D. A. Russell. There, a "'grout' or concrete" building was constructed as an officers' mess hall. Within two years the building was being used for courts-martial and school, which probably means that its construction was not suitable for its original purpose.[36]

Concrete was not a common building material at this time, but it was well known. One of its strongest advocates was Orson S. Fowler, better remembered for popularizing octagonal houses in a book published in 1848. In the revised edition of that book, issued in 1853, Fowler touted the virtues of the "gravel wall," composed of stone, sand, and lime. "Nature's building material is abundant everywhere, cheap, durable, and complete throughout." Fowler's book went through several reprintings and was widely cited. Fowler did not mention cement, but his material was similar to the army's "lime-grout" walls. The army experimented with concrete for seacoast fortifications, and in 1863 Quincy Gillmore of the Engineer Department published *A Practical Treatise on Limes, Hydraulic Cements, and Mortars*, which went through eleven editions.[37] Rogers and Hobbs may have been drawing on either or both of these published works.

Word of Rogers's success with concrete spread even before the Quartermaster General's Office disseminated the information. In November 1872 the post surgeon at Fort Laramie proposed construction of a new hospital to be built of "'grout' or concrete." Lt. Gen. Philip Sheridan endorsed the proposal, mentioning the use of concrete at Sidney Barracks. In January 1873 Lieutenant Rogers sent an explanation of his concrete construction to Maj. Alexander Perry, chief quartermaster of the Department of the Platte, which he then forwarded to Fort Laramie. Construction started at Fort Laramie in June 1873, superintended by none other than Mr. Hobbs. By October 31, the walls were up and Hobbs was discharged.[38]

An important precedent had been set, for the next summer construction began on a larger concrete building at Fort Laramie, a two-story barracks to serve two companies. The cost savings was a deciding factor; the post quartermaster, Charles H. Warrens, estimated that a frame, adobe-lined barracks would cost $8,251.18, while a concrete barracks would cost $6,111.70. Warrens requested permission to hire five stone and brick masons to supervise. Apparently the work went well, because the walls were up by the end of August, and the building still stands today.[39]

Contracts to supply lime and stone for the construction of the concrete hospital went to Adolph Cuny and Jules Ecoffey, civilians who operated a variety of businesses on the edge of the post.[40] One of these was a saloon and way station known as Three Mile Ranch, located on the southwestern edge of the military reservation. At about the same time that they were supplying materials for concrete construction to the army, they built their own concrete building, apparently to house prostitutes. The shed roof and linear plan recalled laundresses' quarters at the fort built twenty years earlier.

Fort Laramie's experience with concrete was positive, as the material was used in the subsequent construction of all of the major buildings at the post: guard house and officers' quarters in 1876; three sets of officers' quarters in 1881; commissary storehouse, bakery, and quarters for noncommissioned officers in 1884; and officers' quarters and an administration building in 1885 (fig. 4-16). The concrete sawmill erected in 1887 proved to be the last structure built before the army abandoned the post three years later. The buildings of Fort Hartsuff in central Nebraska were all built of concrete in 1874–1875; Hartsuff was the only fort in the Department of the Platte to be designed and constructed entirely of concrete.[41]

Fig. 4-16. Fort Laramie, 1889. Several concrete buildings are shown in this village-like view. In the right foreground, laundresses' quarters stand behind a line of laundry; the two-story building behind is the two-company barracks; behind the barracks stands the NCO quarters. Left of these, on the hill, is the hospital, and far left of that, with smoke rising from the chimney, is the steam-powered sawmill. Courtesy American Heritage Center, University of Wyoming.

Although concrete dwellings may have been damp and cold, few such complaints appear in the written record. Perhaps the imperviousness of the walls to strong Wyoming winds—in contrast to porous wood-frame buildings—outweighed any drawbacks of the material. Plaster covered interior walls, sometimes over lath; it was painted or, more rarely, papered. High-traffic areas, such as the stairwell walls in the concrete barracks at Fort Laramie, were covered with tongue-and-groove boards. On the exterior, the concrete buildings were plastered, then scored to resemble ashlar. Despite this effort, concrete looked too utilitarian to win praise as an elegant, or even desirable, material from an aesthetic standpoint.[42]

In 1886 at Fort Robinson, Nebraska, the walls of the concrete hospital, built the year before, began to bow outward. In the subsequent investigation the civilian contractor, a Mr. B. Elliott, cited his experience constructing concrete buildings at Fort Laramie. The investigating board concluded that the concrete probably had been poured too late in the season, and that freezing temperatures had interfered with the setting of the concrete. Inadequate foundations and too heavy a mansard roof were also cited as possible causes. Post Surgeon Walter Reed raised the possibility that there was not enough cement in the mixture. Whatever the exact cause of the problem, concrete as a building material fell out of favor in the late 1880s. When the post quartermaster at Fort Bridger inquired about concrete for a new hospital in 1888, Col. Thomas A. McParlin, medical director of the Department of the Platte, pointed to the Fort Robinson hospital and concluded that concrete was not suited for the climate, as "the cause of the injury to the walls was due to action of frost."[43]

Concrete construction remained an interesting chapter in western fort building, spanning less than two decades. The flow of information was revealing: from practitioners in the field, up to the Quartermaster General's Office, then down to the field in the form of a pamphlet. The period of experimentation with concrete as a building material was also one of technological investigation, indicating that the army was willing to consider innovative solutions to architectural problems.

Model Plans

Despite the efforts of the Quartermaster General's Office to facilitate the construction process in remote regions, that office had to walk a thin line between advice and dictatorship. Post commanders uniformly resented interference with what they considered to be their prerogatives, and post construction was one of these. Although the Quartermaster General's Office gingerly issued model plans in 1872, they were taken more as suggestions than as blueprints. Line officers did not recognize the quartermaster general as having expertise that could benefit them. On the other hand, when the Surgeon General's Office issued model plans for post hospitals, line officers used them much more literally, granting an expertise to the Surgeon General's Office that they did not concede to the Quartermaster General's Office.

Another aspect of the story of standardized architecture is that it helps explain what forts were not. Forts in the nineteenth century were not the same from place to place; unlike today, they did not have identical buildings. Forts did share a certain design sense, particularly in their plan and arrangement of buildings, and designs of some building types appeared at multiple posts. In general, though, standardized architecture—common plans issued from a central source, most logically the Quartermaster General's Office—did not catch on until late in the nineteenth century. Standardized architecture had much to offer the army in the West, besides an image of unity and uniformity. If simply designed, standard plans could provide better spatial arrangements and reduce the cost of building. And if accompanied by specifications and lists of building materials, they could simplify contracting for construction and materials procurement. Since standard buildings were easier to construct, they could also improve the quality of construction. Overall, standard plans should decrease costs, or at least render them predictable. The efficiency inherent in repetitive construction had already been proven, as rows of identical officers' quarters or barracks stood at various western posts. But the same designs were not used at more than one post, since this would have required a central office exercising professional control.

Standard plans faced many practical obstacles and had few proponents. After a few weak attempts and much discussion in the 1860s, in 1872 the Quartermaster General's Office issued the standard plans that had the most impact on military buildings in the West. Ironically, the idea did not originate with the Quartermaster General's Office, and it complied only reluctantly. Its half-hearted promotion of the plans resulted in their being treated more as "model" plans—a suggestion of a design, something to adapt and elaborate—than as "standard" plans to be adopted in their entirety.

Inspector General Randolph B. Marcy headed a panel to revise army regulations in 1871–1872. In soliciting opinions about military issues, the panel received suggested model plans for barracks from Capt. J. M. Robertson. They were routed through General Sherman, who commented: "But the truth is all efforts to prescribe patterns for barracks have failed and must fail, because very few barracks are constructed out and out, but grow—beginning with huts, of earth, stone, logs, board, and even cloth, and ending according to a thousand chances that cannot be foreseen. Fashion too changes, and I rather think local causes must always govern." The proposed plans were also sent to Quartermaster General Meigs who, surprisingly, agreed with Sherman: "The question attempted to be solved is too large for any single solution. Situation, local material, ease of access, and of transportation, permanence of use, and means granted by the Legislative powers are some of the conditions of the problem. . . . I do not advise the adoption of these or of any one plan as the invariable model of barracks. It will be found impossible in practice to apply one plan to all places, climates, seasons, and appropriations." Nonetheless, the Marcy Board recommended strict adherence to established plans, and Meigs complied by producing model plans of six building types. Published in the secretary of war's annual report for 1872, they were introduced in a perfunctory note from Meigs: "The accompanying drawings of military buildings were recommended to the Secretary of War by the Board on Revision of the Army Regulations."[44] He did not include specifications, lists of materials, or any text at all. Most of the residential buildings had halls, porches, gable roofs, six-over-six-light windows, and minimal decoration (figs. 4-17, 4-18, 4-19).

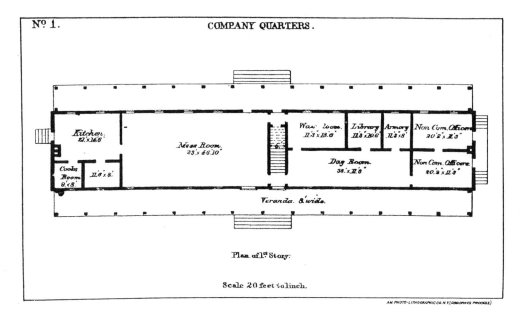

Fig. 4-17. Company quarters, first-floor plan, 1872. Quartermaster Gen. Montgomery Meigs produced plans for barracks based on his 1864 prototype (see p. 42). Courtesy National Archives.

Despite Meigs's reservations, his model plans received wide distribution and served as the basis for a great deal of subsequent military construction in the West. Post quartermasters took from them what they needed most: the plan, especially for officers' quarters. Although the model plans did not adhere strictly to the room allocations, they had the imprimatur of the quartermaster general, and the inspector general usually approved them. To stretch the room allowances, Meigs included additional rooms as "attic" spaces, not counted in the room tallies. Post quartermasters adopted his convention, providing additional, uncounted rooms in the attics.

A typical adaptation of the model plans is seen in double officers' quarters at Fort Laramie. Post quartermaster Charles H. Warrens sent a request to build new quarters up the chain of command in 1873. Meigs advised the use of his "printed model plans," and also recommended concrete (fig. 4-20). Warrens took his advice, but interpreted the plans fairly freely. He kept the basic layout of the plans: halls and stairways along the party wall, parlor and dining room on the outside wall (fig. 4-21). He enclosed an open passageway between the dining room and the kitchen. He also omitted some dormer windows, lowered ceiling heights, and changed proportions of openings.

In 1880 another post quartermaster at Fort Laramie, West Point graduate William P. Hall, consulted the same model plans when he produced drawings for another set of double officers' quarters. The plans that he sent, which were redrawn at the Department of the Platte, included a

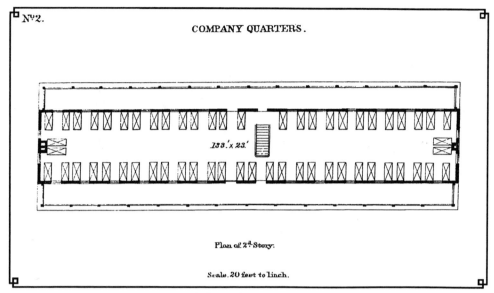

Fig. 4-18. Company quarters, second-floor plan, 1872. The barracks was spacious compared to the earlier model, with fifty-six men sleeping in a dormitory with a ceiling height of 10 feet, which yielded about 550 cubic feet of space per man. Courtesy National Archives.

gable roof, but had no dormer windows and no rear wing (fig. 4-22). As built the next year, however, the two sets of quarters had gained mansard roofs, making them look markedly different from Meigs's version.

Meigs's model plans for commanding officer's quarters served as the basis for those at Fort Bridger (fig. 4-23). In 1883 Chief Quartermaster George B. Dandy of the Department of the Platte submitted plans similar to Meigs's: a center-hall plan with two rooms on either side, fireplaces in the outside walls, and closets in the two first-floor rooms designated as bedrooms (fig. 4-24). In elevation, the house was quite different: unlike the dormered half-story of the model plans, Dandy provided for a full second story with three windows, and a single gabled dormer in the attic. Officials in Washington approved Dandy's plans, which were further transformed during construction. Fancier windows adorned the façade, paired brackets decorated the eaves, and a bay window projected from a room that Dandy had designated a first-floor bedroom (fig. 4-25). Bay windows, which also adorned the two double officers' quarters erected at this time, seem to have had a special cachet. Elizabeth Custer described the construction of new commanding officer's quarters at Fort Abraham Lincoln: "When the quartermaster gave the order for a bay-window, to please me, I was really grateful. The window not only broke the long line of the parlor wall, but varied the severe outlines of the usual type of army quarters."[45] The commanding officer at Fort Bridger in 1883–1884 was Lt. Col. Thomas M. Anderson, who had written a perceptive article about army accommodations the year before.

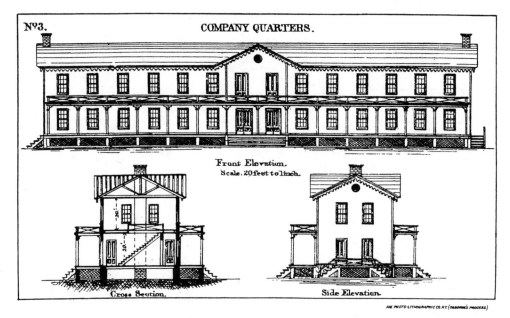

Fig. 4-19. Company quarters, elevations, 1872. The exterior was fancier than the earlier model as well, with a cross-gabled entrance bay, two-story porches, and scalloping at the eaves. Courtesy National Archives.

That the commanding officer's house departed so radically from approved plans was apparently never questioned; any additional costs for what was estimated as an expenditure of $5,310.75 were hidden in a burst of construction that occurred at the post when it received a windfall of $33,500 for construction and repairs in one year.

It was Lieutenant Colonel Anderson, in fact, who pointed out one of the great failures of Meigs's model plans, or of any standard plans at that time. Costs of materials varied so greatly throughout the West that standard plans did not help determine costs with any certainty. Anderson reported that company quarters built on the model plans cost between $7,000 and $18,000; commanding officer's quarters, between $3,500 and $8,000; and double sets of officers' quarters, between $5,000 and $9,000. Anderson concluded that standard plans for officers' quarters were not feasible, due to the "great diversity of opinion." But he explained one rationale for standardization: Brigadier General Pope suggested that rooms in officers' quarters should all be of the same dimensions, with windows in the same places, so that an officer's furniture and carpets would fit in any quarters to which he might be assigned.[46]

Despite protestations, standard architecture was feasible, as was shown by its success with nonresidential buildings. Officers at a post were reluctant to cede control over the design of buildings that they might live in, but were much less territorial about other kinds of structures. Meigs's 1872

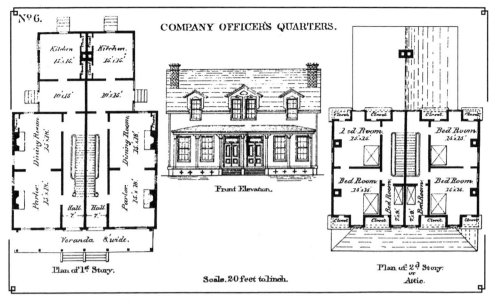

Fig. 4-20. Company officer's quarters, plans and elevation, 1872. Meigs's model plans for double officers' quarters included this symmetrical design with a gable roof. Courtesy National Archives.

model plans included a proposal for a guard house, a rather grand, 98-by-30-foot stone building with a sally port through the middle. A guard house had two purposes: to provide a location for the guard of the day, and to house prisoners. Accordingly, Meigs's plans provided for a large "prison room," or group cell, six individual cells, and a "guard room."

Less than two years later, Meigs issued alternate plans for a more affordable guard house (fig. 4-26). About half the size of the previous one, this building had cells independent of the walls, so that the exterior walls of the building could be of any material. The cells were built of "Plank 6" x 2½" and 7'-0" high, grooved and tongued with iron bars 1" x ¼ inches." Above this 7-foot-high wall was a grating of iron bars to the 9-foot-high ceiling. The plan provided for a room for "general prisoners," a room for "casual prisoners," and three solitary cells, as well as an office for the guard. The building had high, barred windows above a row of loopholes for firing out; clearly, Meigs conceived the guard house to have some defensive function. Adjacent to the guard house was the 50-by-50-foot prison yard, "surrounded by a stockade" and containing a privy. This second model plan for a guard house proved more popular than the first. In 1876 Fort Laramie gained a guard house built of concrete, of the same dimensions as the model plans and clearly based on them. Two years later Fort D. A. Russell acquired a new guard house that was also based on the model plans (figs. 4-27 and 4-28).[47]

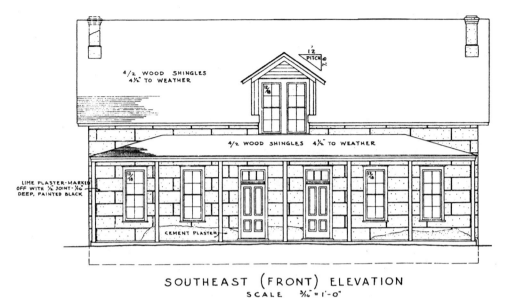

Fig. 4-21. Fort Laramie, double officers' quarters, elevation, delineated by Leslie E. Wilkie, 1939. Post Quartermaster Warrens adapted the model plan (fig. 4-20) in 1875 to produce this design, constructed in concrete scored to resemble stone. Courtesy Library of Congress, HABS.

The 1872 model plans also contained designs for a bake house—a small structure with two brick ovens (fig. 4-29). Although the chief quartermaster of the Department of the Dakota used these plans for bake houses at Forts Meade, Abraham Lincoln, and Keogh, no bake house of this design appeared in the Department of the Platte. In 1882 the commissary general of subsistence issued *Notes on Bread Making, Permanent and Field Ovens, and Bake Houses*, prepared by Maj. George Bell. After detailing information on the selection of flour, making of bread, construction of ovens, and selection of tools, Bell provided plans and elevations of two bake houses with one or two ovens. Bell's plans were used to gain funding for a new bake house at Fort Laramie in 1883, but the completed bakery bears little resemblance to Bell's plan.[48] Nonetheless, the issuance of standard plans by the commissary general of subsistence, taking the initiative away from the quartermaster general, reveals the growing influence of professionalization. Interestingly, army surgeons had claimed their turf even earlier.

Standardization of Hospitals

Hospitals were the first buildings to be effectively standardized, and their uniform appearance at various posts testifies to that success. The surgeon general had more control over hospital designs than the quartermaster

Fig. 4-22. Fort Laramie, double officers' quarters, commanding officer's quarters on left, 1883. The same model plan (fig. 4-20) was used in 1880 for officers' quarters constructed with mansard roofs. Courtesy Wyoming Division of Cultural Resources.

general ever did over other buildings. The surgeon general could cite compelling reasons of health, and his scientific expertise enabled him to make hospital planning the purview of his department. Although doctors were just gaining professional recognition in the mid-nineteenth century, they had higher status, certainly, than architects.

Hospitals occupied an odd niche in American society. Traditionally, women undertook nursing in the home. Only segments of society without homes and women, such as paupers and soldiers, required hospitals. At army posts, the hospital provided places for immediate medical care in the surgeon's office, long-term convalescence in the wards, and storage of medicines in the dispensary. Kitchen and mess room were necessary to feed the patients, rooms for hospital steward or nurse housed attendants, and a post-mortem room for autopsies was often located in a nearby building. The medical emergency most often associated with the army, battle wounds, was actually one of the lesser threats to soldiers' health, statistically. In 1872, for example, the number of deaths per 1,000 men from gunshots, accidents, and injuries was 4, while diseases claimed 9. The national average was 1 and 6, respectively. At Fort Bridger in 1874, out of 186 officers and soldiers, only 1 died, from an unspecified local disease. There were

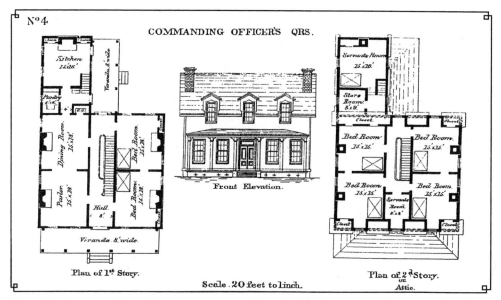

Fig. 4-23. Commanding officer's quarters, plans and elevation, 1872. Meigs's model plans for commanding officer's quarters prescribed a center hall and a kitchen wing. Courtesy National Archives.

37 cases of diarrhea and dysentery, and 42 of catarrh and bronchitis. There were no gunshot wounds, but 32 cases of accident and injury.[49]

Before the Civil War little thought was given to specific hospital designs in the West. In 1870 Fort Laramie's post surgeon, Henry S. Schell, described his hospital and illustrated the haphazard attention that hospitals normally received. The hospital, built of adobe in 1856, "consists of two rooms, each 20 feet square, separated by a hall and dispensary, which communicate with each; the large rooms have an air space of 4,200 feet, and contain eight beds. They are warmed by open fireplaces, and are not ceiled, the roof being of heavy logs, with a covering of shingles; their walls are 20 inches thick." In 1858 a long wing containing a steward's room, ward, storeroom, dining room, and kitchen was constructed of a heavy timber frame infilled with adobe. In the winter of 1866–1867, scurvy plagued the troops, and the hospital was augmented with three hospital tents, used all winter long, which "were repeatedly blown down, and by the time spring came were torn into ribbons." That summer the post surgeon designed an addition on the other side of the steward's room, making a T-shaped plan (fig. 4-30). The adobe-filled frame addition contained a 25-by-55-foot ward with twenty beds, as well as a bathroom and watercloset.[50]

Fort Bridger's hospital was also poorly planned. The original building, 113 by 18 feet, contained two wards, a dispensary, bath- and washrooms, and the attendant's room. A 62-by-20-foot wing at right angles, forming an

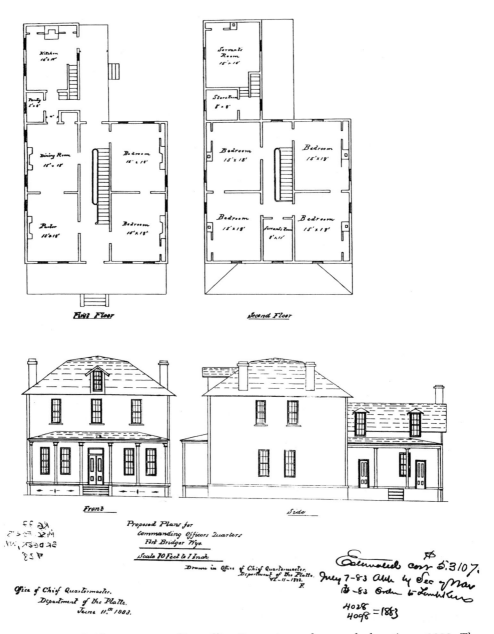

Fig. 4-24. Fort Bridger, commanding officer's quarters, plans and elevations, 1883. The chief quartermaster of the Department of the Platte adapted the drawings (fig. 4-23) eleven years later. He raised the roof to create a full second story, but retained the plan. Courtesy National Archives.

L-shaped plan, held a steward's room, storeroom, dining room, and kitchen. Constructed of logs, the building had only 7-foot ceilings in the wards. The post surgeon complained that "the building is badly adapted to the purpose for which it was designed. The medical officer on duty at the time it was

Fig. 4-25. Fort Bridger, commanding officer's quarters, 1884. When built, the house gained a bay window, decorative hoodmolds, and paired brackets at the eaves. Courtesy Wyoming State Archives, Department of State Parks and Cultural Resources.

erected disclaims all responsibility for its bad design and says his opinions were ignored by the quartermaster who constructed the building."[51]

The Surgeon General's Office disapproved of these unplanned hospitals; standard designs offered a means of achieving uniform quality (fig. 4-31). First Lt. Don Carlos Buell's plans, published in 1861, contained one for a hospital that was severely criticized by the Surgeon General's Office. Assistant surgeon John Shaw Billings, in his comprehensive "Report on the Barracks and Hospitals of the United States Army" in 1870, wrote of Buell's design: "the small wards, shut off from the open air on one side by a hall, the deficient air space, and the presence of the water-closet in one corner of the ward, so to speak, are all in opposition to the first principles of correct hospital construction."[52] Billings's interest in ventilation and sanitation, reflected in this critique, guided much of army hospital construction in the late nineteenth century.

Billings himself was a fascinating character. Pursuing his concern for ventilation in army buildings, he became an expert on the subject, publishing articles on ventilation and heating in *The Sanitary Engineer* in the early 1880s, followed by two books on the same subject. One of his books

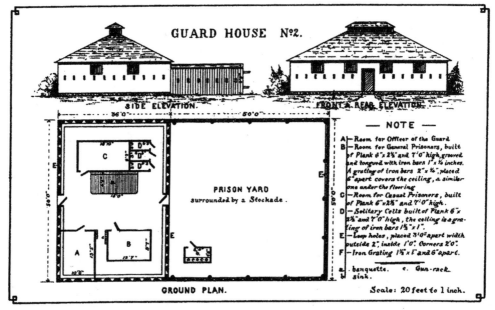

Fig. 4-26. Guard house, plan and elevations, 1874. Meigs's 1872 model plans for a guard house were too elaborate and were predicated on stone construction, so two years later he issued this cheaper version. Courtesy National Archives.

was still being cited thirty years later. He also maintained his interest in architecture, producing an essay on "Hospital Construction and Organization" for Johns Hopkins Hospital, which then hired him to design its new hospital building, completed in 1889. He served as president of the American Public Health Association in 1880. After retiring from the army, he headed the New York Public Library, and oversaw completion of its new building to his sketch plans. His interest in libraries arose from his experience in directing the evolution of the library of the Surgeon General's Office into the Army Medical Library, for which he compiled a comprehensive bibliography of all medical publications. His supervision of the collection of vital and medical statistics in three censuses—1880, 1890, and 1900—demonstrated his interest in vital statistics.[53]

For the army, Billings wrote two comprehensive reports on barracks, quarters, and hospitals in 1870 and 1875. Both of his publications, issued as circulars by the Surgeon General's Office, included compilations of reports from surgeons stationed at every post, describing the location and architecture of each post as well as the general health of the men. Billings wrote lengthy introductions to both reports that commented on the architecture, with particular regard to ventilation and sanitation. Although the reports contain valuable descriptions and recommendations, Billings spent his entire career in the East and apparently never visited any army posts in the West.[54]

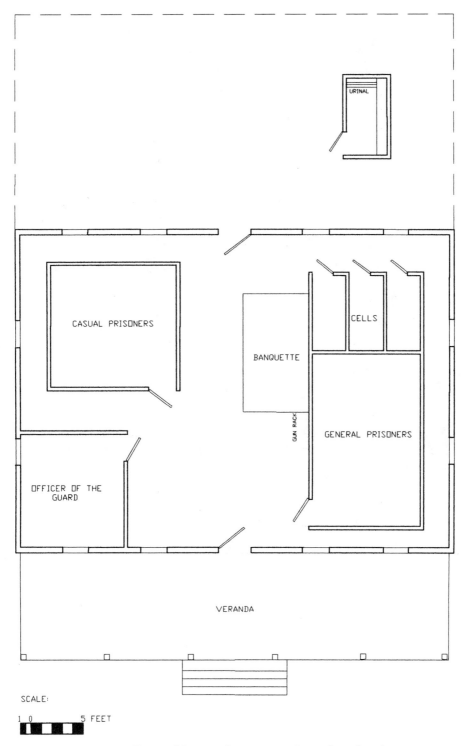

Fig. 4-27. Fort D. A. Russell, guard house plan, 1878. Indicated on the plan are two group cells and three individual cells, with a privy in the fenced yard. From National Archives, redrawn by author.

Fig. 4-28. Fort D. A. Russell, guard house, ca. 1895. Eleven years after construction, the guard house was doubled in size with lateral wings. Courtesy National Archives.

Billings's thinking on these matters reflected current medical practice. In the mid-nineteenth century, scientists believed that ventilation was necessary to replace consumed oxygen, remove carbon dioxide (also called carbonic acid, which they perceived as much more detrimental than we do today), and remove organic contaminants in exhalation. Experts including Assistant Surgeon Billings believed that exhalation was "contaminated with organic matter which has a strong tendency to putrescence, and has been well described as a sort of 'aerial filth;' or, . . . 'a physiological miasm,' which is directly and positively hurtful when introduced into the system." This organic matter originated not only in exhalation, but with the entire body; he pointed to "air contaminated with organic products thrown off by the lungs and skin."[55] Not yet adopting the germ theory, these scientists perceived that contagion occurred through proximity. Rather

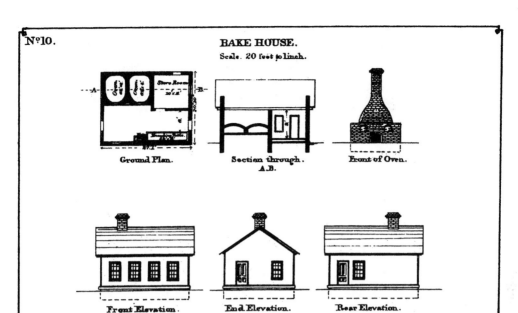

Fig. 4-29. Bake house, plan and elevations, 1872. The bakery was an important building for soldiers' subsistence. Post quartermasters used Meigs's model plans at other posts, but not in the Department of the Platte. Courtesy National Archives.

than attribute it solely to germs carried through exhalation or bodily fluids, they understood it to be attributable also to emanations from the body.

Ventilation prevented diseases transmitted by this bodily miasm. Natural cross-ventilation was easily achieved in hospitals and barracks having a long narrow dormitory with windows on both of the long sides. Accordingly, Billings eschewed plans with squarish, small rooms, particularly those with "dead walls," or rooms without cross-ventilation. Instead, he advised long narrow plans, never more than 24 feet wide. Windows, provided at a rate of one for every two beds, should have a double-hung sash, so air could enter at top and bottom. Windows were often augmented by ventilation shafts in the roof ridge. And because the building itself was believed to become imbued with these poisons and emanations, Billings recommended the replacement of hospitals every fifteen years. He found wood a more suitable material for such a short-lived building than stone or brick. For both barracks and hospitals, Billings recommended the use of a "ventilating double fireplace" which incorporated flues for fresh air (fig. 4-32).[56]

In his 1870 report on barracks and hospitals, Billings advocated the pavilion plan of hospital, in which pavilions containing the wards were separate from the offices, kitchen, and laundry. The pavilion plan was first

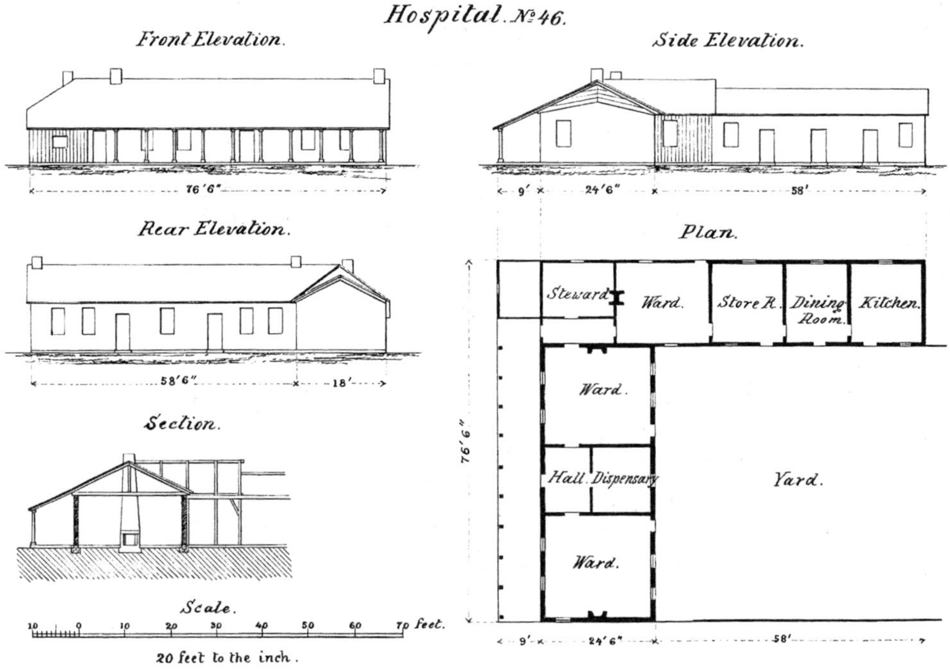

Fig. 4-30. Fort Laramie, hospital, plan and elevations, 1867. The portion with two wards was built of adobe in 1856. Two years later, a frame wing was added. Soon after this drawing was made, another wing was added, making a T-shaped plan. Courtesy National Archives.

promoted by the French Académie des Sciences, which issued an extensive report on hospital design in 1786; by this time it was gaining popularity in the United States. Billings praised the standard hospital plan issued by the Surgeon General's Office in 1867, which had a two-story main block, containing offices and kitchen, and two lateral, one-story wings for the wards. Well-positioned windows provided cross-ventilation, supplemented by ridge vents in warm weather, and by shafts related to the stove in cold weather. Rooms at the end of each ward housed water closets if running water was available; if not, these rooms contained "air-tight close stools," to be emptied frequently, with additional sinks in the yard. This standard plan allowed porches around the whole building when hospitals were constructed in the south. The hospitals could be constructed of wood, brick, or adobe.[57]

Fort D. A. Russell's post surgeon, C. H. Alden, cited these standard plans when he described the post hospital in 1870. He claimed that the hospital "essentially" followed the standard plan, but then mentioned some changes he had made. Because Fort D. A. Russell was north of 38°N, the hospital was built without porches. Accordingly, the front hall was widened from 5 feet to 6 feet to accommodate "those presenting themselves at surgeon's call."

THE HOMELIKE APPEARANCE OF THE HOUSE ■ 163

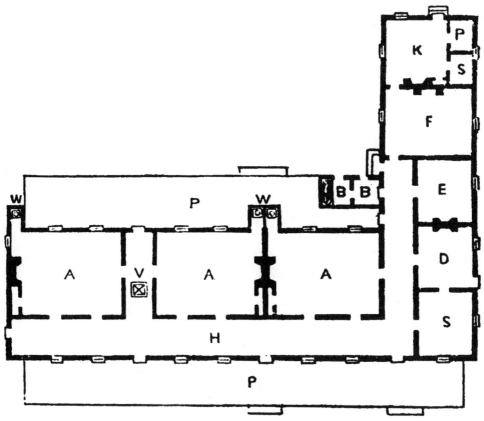

Fig. 4-31. Hospital, plan, 1870. Issued in 1860 as a model plan, it was criticized by the assistant surgeon general because the wards were not cross-ventilated. From U.S. War Department, Surgeon-General's Office, Circular No. 4, *A Report on Barracks and Hospitals, with Descriptions of Military Posts*, xxv.

A skylight was added in the stairhall, a linen closet added at one end of the second-story hall, and a post-mortem room added in a separate building, rather than in a room on the second floor. The ventilation system proved unnecessary, due to "the severity of the winds, and also the low range of temperature, particularly at night." Air channels in the floors of the wards were constructed but never opened; ventilators in the ridge were opened only in July and August. The hospital was not used entirely as a hospital; one ward served the purpose for which it was intended, but the other was used as a chapel.[58]

The application of standard plans was uneven, not because post quartermasters had better ideas of how to build hospitals, but because funds and materials approved for hospitals were often diverted to other uses. To ameliorate this situation, in 1870 the adjutant general issued General Order No. 118, which gave the surgeon general greater authority over hospital

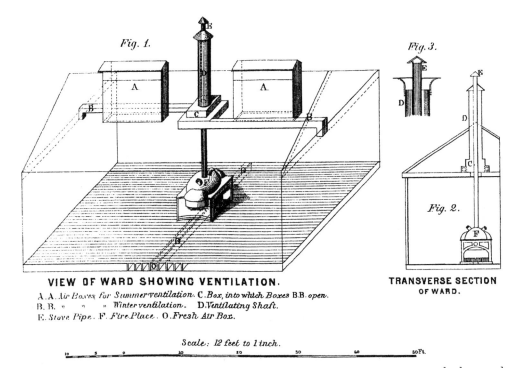

Fig. 4-32. View of ward showing ventilation, 1871. The assistant surgeon general advocated use of the ventilating double fireplace, which brought fresh air in through ducts under the floor. From Surgeon General's Office, Circular No. 2, 27 July 1871, National Archives, RG 112, Entry 63, Box 3, plate VI.

construction. Beginning in 1872, an annual appropriation of $100,000 was devoted to hospitals, to be spent at the discretion of the surgeon general. Although the post surgeon still needed myriad approvals and the cooperation of the commanding officer and post quartermaster to build or repair a hospital, the involvement of the surgeon general and his department- and division-level subordinates gave the post surgeon additional authority. With this further approval required, post quartermasters preferred to use the standard plans of the Surgeon General's Office. In 1888, for example, Post Quartermaster Alexander Ogle at Fort Bridger, anticipating a new structure, requested a copy of the "approved plans for hospitals" from the medical director of the Platte.[59]

General Order No. 118, issued by the Adjutant General's Office, also noted that "plans and specifications for post hospitals approved this day will form the basis of action." These plans and specifications, issued by the Surgeon General's Office in 1870 as Circular No. 3, provided plans for a hospital of twenty-four beds very similar to those issued in 1867. Allowances were made for detached wards and kitchens in hot climates. As Billings noted a week after issuing these plans, "It is manifestly impracticable to plan a

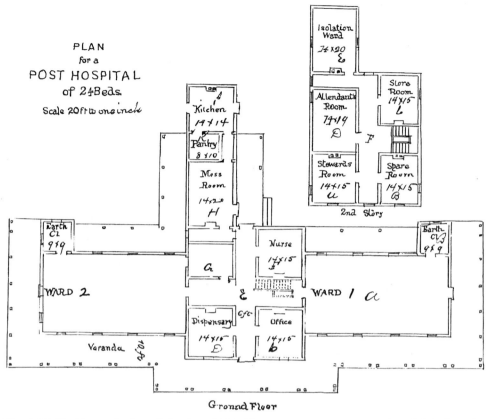

Fig. 4-33. Fort Laramie, hospital, plan, 1874. The post quartermaster enclosed this drawing in a letter indicating completion of construction thus far. The walls were up and roof laid on all the lettered rooms, but the veranda was not added until 1878, and ward 2 was never built. Courtesy National Archives.

hospital which shall be equally suited to the burning mesas of Arizona and to the bleak North Atlantic coast." The next year, the Surgeon General's Office issued new plans on the same model that included estimates of materials and gave specifications detailed enough to incorporate into a contract.[60]

The standard plans of 1871 were the source for the hospital built at Fort Laramie. The post surgeon, Robert M. O'Reilly, could not resist proposing several changes. He suggested converting the 9-by 9-foot earth closets at the end of each ward into 12-by-12-foot bathrooms, containing a tub, an earth closet, and a "lavatory," with an opening to the outside as well as to the ward. This was apparently not done, but the space originally designated as a bathroom (G on 1874 plan) was converted into additional storage space "as the post receives yearly supplies of medical stores" (fig. 4-33). O'Reilly also provided for indoor access to the mess room from the wards; the original plan required patients to go outdoors. O'Reilly urged concrete construction, but required furring strips inserted into the concrete to receive laths, which would provide "sufficient air space for the maintenance of an even temperature through the hospital."[61]

The request for new construction went forward in 1872, and building commenced in the summer of 1873. The concrete walls went up quickly, but the hospital was never completed; one of the two wards was never built. By the spring of 1874 the walls were up, the floors laid, and the building roofed. The hospital needed plastering, windows and doors, stairs, and porches. The porches were important. As the post surgeon observed in 1873, they were necessary to protect windows from hailstorms, as "such a storm recently destroyed 800 panes of glass in the post." Shutters were necessary for windows not protected by porches.[62] The porches were added in 1878.

In 1875, Billings again reviewed hospital construction throughout the military and found fault with some of his earlier pronouncements. "The 'pavilion plan,' which for a time was supposed to be a perfect panacea against all evil, has been found, by sad experience, to furnish no security against the evils summed up in the word 'hospitalism.'" Convinced that poisonous air would seep upward, Billings still recommended that wards not be located beneath any other part of the hospital. Despite his disillusionment with the pavilion plan, Billings used it in his design for the widely publicized Johns Hopkins hospital. George B. Post's New York Hospital (1875), a high-rise building that proposed a new model based on antiseptic surgery and mechanical ventilation, remained an isolated example.[63]

In 1877 the Surgeon General's Office issued new plans and specifications, yet again based on the same general model. When Fort D. A. Russell built its new brick hospital in 1887, it turned to these 1877 standard plans, but made some adaptations, such as the hip roofs on the lateral wings (figs. 4-34 and 4-35). The availability of running water meant that the building soon needed an addition, however, and in 1891 local contractor Moses P. Keefe built two wings, forming an E-shaped plan. Each wing contained a bathroom with tub, two water closets, one urinal, and three basins, as well as a storeroom.[64]

The success of the Surgeon General's Office with standard plans is reflected in the common appearance of hospitals at posts throughout the West. Fort Laramie's concrete hospital of 1873–1874, though never completed, looks very similar to Fort D. A. Russell's brick hospital of 1887: both had a two-story main block, one-story lateral wings, and porches around the buildings. Also, the hospital's location away from the parade ground, usually on a high, "healthful" site, added to its recognizable presence on the landscape.

THE HOMELIKE APPEARANCE OF THE HOUSE ■ 167

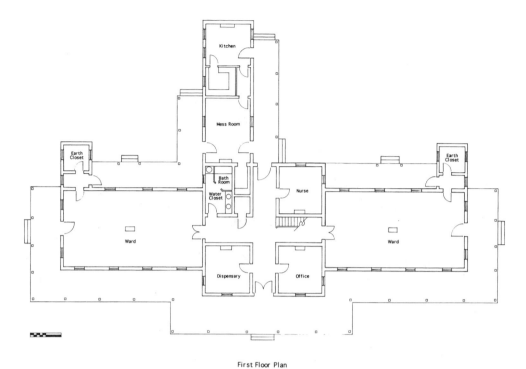

First Floor Plan

Fig. 4-34. Fort D. A. Russell, hospital, first-floor plan. The surgeon general's model plans of 1877 were followed quite closely for the hospital at Fort D. A. Russell built ten years later. Drawing by Julie Osborne, 1996, Western Regional Architecture Program, Graduate School of Architecture, University of Utah.

The surgeon general's success in enforcing standard hospital plans contrasts sharply with the quartermaster general's failure to standardize other buildings. The surgeon general faced the same problems as the quartermaster general: available building materials varied from post to post; no one design proved suitable for every location; costs ranged widely; and post quartermasters and surgeons would always make their own modifications. Nonetheless, the surgeon general's standard plans were far more widely accepted than the quartermaster general's. Other military buildings continued to vary from post to post.

Need for Standardization

Not until the turn of the century did standardization gain a better reception among line officers. Once troops were relieved of construction tasks and post quartermasters contracted out the work, the advantage of flexibility inherent in post-designed architecture was lost. After regulations mandated

Fig. 4-35. Fort D. A. Russell, hospital, ca. 1895. Men cluster on the shady side of the hospital in this view, showing the utility of the extensive porches. Courtesy National Archives.

that construction be contracted out, Forts Bridger and D. A. Russell received new brick and stone buildings in the 1880s. In 1886 the post quartermaster at Fort Bridger, 1st Lt. Francis Eltonhead, received authorization to contract out the construction of a new barracks, guard house, and oil house, and to build them of stone. Unfortunately, the West Point–educated Eltonhead was not up to the task, which involved advertising for and contracting with suppliers of building materials as well as a building contractor. He was forced to ask Chief Quartermaster George B. Dandy in Omaha how to calculate the amount of stone needed: "that is, . . . the number of perches of stone building 'measured in wall.' For example, a building is 20 by 20 feet outside measurement and 14 feet high, walls 18 inches thick. How many perches of stone are there in the building?" Apparently, he received no answer, because three weeks later he posed the same question to two building contractors, and added, "is it customary to measure the corner twice?"[65]

Meanwhile, Eltonhead's specifications did not match his plans. The foreman at the site, a Mr. Clement, noticed that the advertisement called for a 30-by-31½-foot guard house of outside measurement, but the plan indicated that these were inside measurements. Eltonhead assured the chief quartermaster that there was no misunderstanding between Mr. Clement and himself, but Clement apparently wanted it clarified with his employer, George H. Jerrett from Fort Duchesne, Utah. A month later, Eltonhead confessed that "another mistake was made" in regard to the barracks. The plan indicated a rear wing 33 feet wide, although the advertisement said

30 feet. Clement's solution was to extend the length in order to get the same square footage, but Eltonhead needed the contractor's approval.[66] These errors would not have been noticed if soldiers had been constructing the buildings; the discrepancies would have been fudged and a compromise reached on site. But once post quartermasters were held to their plans and specifications by contract, a higher order of accuracy was required.

A similar fiasco occurred when 1st Lt. Edward Chynoweth, post quartermaster at Fort D. A. Russell, produced a design and issued a contract for two field officers' quarters in 1887 (fig. 4-36). Charles F. Humphrey, the depot quartermaster called in to supervise construction, found fault with the plans, stating that the plans and specifications developed "a great want of harmony." He called the plans "meagre, defective, and *not drawn to scale.*" More specifically, he identified the flues as inadequate, the hall unheated, the front stairs poorly designed, the back stairs unmanageable and unlighted, and the one bathroom poorly placed, opening into a bedroom. Additional inadequacy of Chynoweth's plans was due to plumbing and sewerage; the full bathroom and hot and cold running water called for in the specifications did not appear in the drawings (fig. 4-37). Humphrey produced new plans of a more elegant design, with sliding doors and a more graceful staircase. He moved the bathroom to the first floor and delineated full plumbing.[67]

The difficulty was that buildings had become more complicated since Chynoweth had graduated from West Point in 1873. Although Quartermaster General Holabird supplied every post with a subscription to *The Sanitary Engineer*, complex plumbing designs were generally beyond the capability of someone trained as an officer, not as a construction supervisor or an architect. An anonymous staff officer concurred, praising Quartermaster General Holabird: "It is a fact not to be disputed that the present Quartermaster General of the Army is far better versed than any of his predecessors in the very important subjects of building, water supply, sewerage, etc. . . . all the officers of his Department have been supplied with standard works on architecture, building, drainage, sewerage, etc., as has for many years been done in the Medical Department and Engineer and Ordnance Corps. Formerly nothing was furnished them."[68]

The quality of army architecture was still the object of many complaints in the 1880s. In 1887 the *Army and Navy Journal* ran a series of architectural plans and perspectives provided by the Cooperative Building Plan Association, published in *Shoppell's Modern Houses*. The *Journal* called them "designs for officer's quarters," hoping to elevate the general quality of design at posts. The plans were elaborate, with bay windows, servants' stairways, bathrooms, and asymmetrical arrangements of rooms, while the perspectives showed a variety of shingles and clapboards, leaded glass,

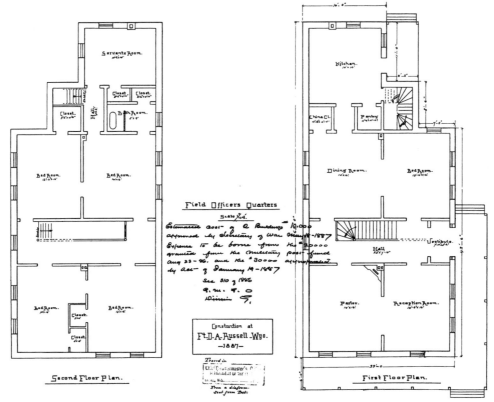

Fig. 4-36. Fort Laramie, field officer's quarters, plans, 1887. Post Quartermaster Edward Chynoweth produced these plans, which Depot Quartermaster Charles F. Humphrey called "meagre, defective, and *not drawn to scale*." Courtesy National Archives.

and irregular roofs. At least one reader was hopeful: "You are doing a great, good work for the Army by publishing designs for Officers' Quarters, as it will open the eyes of the Quartermaster's Department to the fact that a picturesque and convenient house can be built as cheaply as have been the ugly and badly arranged ones at the posts, without furnaces, laundries, closets, or other conveniences." A staff officer agreed: "It is hoped that the agitation of this subject may result in some improvement in the present architecture (?) of our large garrisons." He noted that buildings continued to be designed at posts, and charged that "the result is that the class of work now being done at posts is not in advance of that done 10 years ago."[69]

With these recognized needs—better designs, complex plumbing and heating, and professionally developed specifications—standardization would find a new acceptability at the turn of the century. Officers in the field conceded that they did not have the skill to satisfy these needs, and that professionals in Washington were better suited to the task. While line

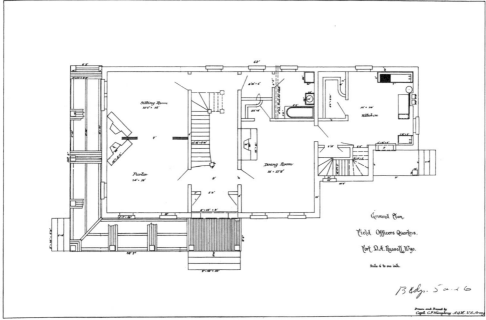

Fig. 4-37. Fort Laramie, field officer's quarters, first-floor plan, 1888. Depot Quartermaster Humphrey then produced his own plans, also with the main entrance on the side. Courtesy National Archives.

officers lost the flexibility that they had when they retained control of design and construction, they gained a smoother and more efficient building process when they turned design over to standard plans developed in Washington, construction supervision over to professionals, and construction over to contractors.

While by 1890 the stage was set for standard architecture army-wide, it was not yet a reality. Western posts continued to exhibit a variety of architecture more akin to individual development than to a construction project planned by a large government agency. There were trends toward modernization: the adoption of hospital designs, the wider range of building materials resulting from innovation or brought in by the railroad, and the move to contract out construction. But these were outweighed by factors persisting from the antebellum era: a desire to retain design control at the post or department level, inadequate funding resulting in dilapidated buildings, and unchanging regulations as to space. Buildings continued to reflect genteel designs from the eastern establishment, but often with picturesque touches popular back East, such as bay windows, bracketed eaves, and hoodmolds. The furnishings of the residential buildings and the conveniences available at the posts also revealed the not-quite-modern aspect of western forts.

5

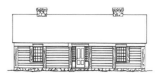

QUITE AN AIR OF HOME COMFORT

Furnishings and Conveniences

In 1878 the *Army and Navy Journal* assured readers that there was culture out West: "Here and there men of intelligence and enterprise, and women of beauty and education, have formed nuclei of solid cultured society. Especially recognizable is this superior condition of society in the neighborhood of cities where Government military posts have been established." As evidence of this, the paper described a luncheon given by Carrie Merritt, wife of Wesley Merritt, 5th Cavalry, commanding officer at Fort D. A. Russell. The hostess's credentials were given: Carrie Merritt was "one of the belles of Cincinnati," "one of the liveliest and most intelligent conversationalists in the land; has twice made extended European tours, and is esteemed the most generous of entertainers." Her guests—three officers' wives and four ladies from the town—wore silk, lace, and velvet dresses ornamented with pearl, coral, diamond, and cameo jewelry. The menu included raw oysters, quail, calf's head, chicken salad, and Charlotte russe. Coffee was served in French china of 1796 in an "exquisitely furnished" parlor. After lunch, the regimental band serenaded the ladies.[1]

This luncheon reveals a surprising level of sophistication for a western post (fig. 5-1). Although it took place in an unfashionable, board-and-batten-sided house, the hostess was obviously able to tap into great resources. French bonbons, California pears and apples, and raw oysters traveled by rail to Cheyenne for this December 24 luncheon. The ladies were well traveled, too. Several of them had journeyed abroad, and all had come to Cheyenne

Fig. 5-1. Fort D. A. Russell, commanding officer's quarters, ca. 1867–1868. Built in 1867 at the top of the diamond that was the fort's original parade ground, the wood-frame house served as the commanding officer's quarters until the end of the century. Courtesy Wyoming State Archives, Department of State Parks and Cultural Resources.

from other parts of the country. They dressed fashionably and obviously took delight in creating a formal luncheon that would have been just as appropriate in Boston or New York. And, despite its appearance in the newspaper, the luncheon was not an unusual event. Officers and their wives often traded invitations with the city's upper crust for dinners and dances. On New Year's Day, officers' wives opened their homes to callers from the city, and on summer days Cheyenne's well-heeled citizens took their carriages out to the fort to hear the band play.[2]

The socializing reflected the class standing of officers and their wives; they placed themselves in the upper class and in doing so had certain standards to uphold. The desire to maintain standards of eastern civilization could be seen not only among officers' wives, but also among military leaders on behalf of their soldiers. The army frequently compared its accommodations to civilian standards. In his last annual report, General Sherman said, "The day is past when a soldier will be content to live in 'dug-outs,' on 'his pound of bread, pound of meat, and gill of whisky,' per day, whilst the farmer, mechanic, and laboring man alongside has a good house, with coffee, sugar, vegetables, and a well-provided table." In 1887 the commanding officer of the Department of the Platte, Brigadier General Crook, argued that "there would seem to be no good reason why an officer's quarters, which, while he is in the service, are his only home, should not have the conveniences that are found in the ordinary houses occupied by civilians. No quarters should be constructed which are not provided with bath-rooms."[3]

As far as health and comfort were concerned, army officials attempted to provide the latest conveniences, although with mixed success. Modern ideas of comfort and professionalism affected the provision of furnishings and conveniences. The quartermaster department under Montgomery Meigs developed specifications for certain items that would enhance the men's comfort, including iron bunks in 1872, stoves (1876), barrack chairs (1878), and oil lamps (1881). The debate between ventilation and heating, which became a contest between medical expertise and men's comfort, ended in a draw. Barracks plans changed to facilitate ventilation, but warmer stoves won out over fireplaces. The increased level of comfort, enhanced by furnishings and stoves, reflected changing standards of what was appropriate for soldiers. The provision of more complicated systems, which occurred during Samuel Holabird's term as quartermaster general, was more difficult. Sanitation efforts suffered from a lack of technical expertise. The installation of water tanks, steam engines to run pumps, and piping to provide water to the post required specialized training. Electric lighting was also in the hands of experts. Both plumbing and electricity took years to achieve. While they might have enhanced men's comfort and made the posts modern, the complexity of these systems prevented their rapid installation.

The belated provision of conveniences was consistent with the fort's village character. There is no doubt that the army surpassed many of its rural neighbors in the provision of amenities, but these forts were small settlements, not rural sites. Less sophisticated than modern cities, the western forts did not usually pioneer the application of comforts and conveniences. Fort D. A. Russell's evolving relationship with Cheyenne reveals a shifting balance. Initially, the fort led the town in its establishment of civilized living, having greater resources to build from scratch. But in the 1880s, when Cheyenne gained residents with real wealth, it soon surpassed the underfunded fort in the provision of amenities. This uneven progress to modernity, seen in post quartermasters' reluctant acceptance of standard architecture, as well as in the provision of furnishings, ventilation, sanitation, and illumination to officers and soldiers, contributed to the provincial nature of these forts.

Furnishings

Occupants of military architecture were transient. All were stationed temporarily, ready to move on orders that could come at any time. No one, except the post trader, owned his dwelling. Besides having little incentive to alter their barracks, enlisted men lived too densely together to allow for much personal space. Officers' wives devoted the most effort to individualizing their quarters, despite their temporary residency. Their mission was to civilize their husbands, the troops, the fort—whatever their sphere of influence might be—and the easiest way to express values of domesticity was to domesticate their quarters.

This task of civilizing was not solely a female pursuit, however. Just as the army civilized the West in terms of eliminating the perceived savages, the army also brought its own refinements in terms of domestic life. Whether or not they were accompanied by their wives, officers were expected to live in a civilized manner. Furnishing their quarters was the easiest way to maintain this sense of civilized life (fig. 5-2). Because the freight that officers could carry with them from post to post was limited to 1,000 pounds, personal touches tended to consist of cloths and coverings that were easily transportable. Upholstered furniture, commonly the sign of a well-appointed parlor, was usually too expensive and too bulky for an officer's lifestyle (fig. 5-3). Homemade substitutes had to suffice. Emily FitzGerald, the post surgeon's wife at Fort Boise, Idaho, described how she and her husband constructed a "lounge." Her husband made a frame, and she had "two square hair pillows" and "some red cretonne." "I had Mrs. Burroughs make me a mattress and had it stuffed with hay. Doctor [her husband] tacked it for me just like a bed mattress, and you don't know what a fine lounge we have." So fine, in fact, that others at the post started emulating it. "Indeed, we have set the fashion. There are two on the way being made like it now, and the carpenter came in a few minutes ago and asked if he could see my lounge frame. He is going to work on a third just like it for Mrs. Riley."[4]

> ### "How They Live on the Plains," 1867
>
> An article written from Fort McPherson, Nebraska Territory, in 1867, gives advice to officers coming to the Great Plains:
>
> > In the first place, officers serving on the frontier are in garrison, and have quarters. Sometimes they are very poor, it is true, still, they are furnished with quarters of some kind, and do not generally live in

> tents. . . . These quarters are furnished by the quartermaster, with heating and cooking stoves; with the usual appurtenances, a plain table or two, a rough bunk, or bedstead, and possibly a chair made of pine boards. Everything else that an officer wants, he must provide for himself. . . .
>
> Officers who have been on duty in this country long enough to learn how to live, and who have any taste for living like gentlemen, generally furnish their quarters comfortably, and sometimes, quite elegantly. Curtains are extensively used, and carpets are not altogether unknown, even in bachelor quarters.
>
> They provide themselves with comfortable chairs and, with suitable cloth, and the aid of a company carpenter, and a few pine boards, they make very respectable lounges, and, having hung up a few handsome engravings, often present a very comfortable, and civilized appearance in their "one or two rooms and a kitchen." . . .
>
> Every officer coming on the Plains, should bring with him, *first*, a good mess chest, well furnished for four or six persons; a good roll of bedding—a matrass [sic], a few comfortable chairs (some pattern of camp chairs are the best, as they are easily transported), a trunk filled with a good supply of clothes for at least one year. . . .
>
> Beside these necessary articles, a table spread or two, a few curtains, some cloth suitable for covering furniture, and a roll of carpeting, will not come amiss. Further than this, he will never regret bringing any small articles of luxury that his taste may suggest, for he will find very little difficulty in taking them with him from station to station when he changes from one to another.
>
> In a word, officers coming to the Plains should remember that they are going among gentlemen who live as such—as far as the circumstances for the case will admit—and not among a set of frontier ranchemen [sic], who sleep on the ground and eat fried bacon from their fingers.
>
> A. R.
> Fort McPherson, N.T. December 22, 1867
>
> Source: *Army and Navy Journal* 5 (11 January 1868): 330.

Given that officers, even when alone, were expected to maintain high standards, the expectations for the civilizing influence of wives were even greater. As a newlywed officer's wife at Fort D. A. Russell in 1874, Martha Summerhayes confronted her new quarters, with the addition of her husband's furnishings: "Jack had placed his furnishings (some lace curtains, camp chairs, and a carpet) in the living-room, and there was a forlorn-

Fig. 5-2. Fort Meade, South Dakota, commanding officer's quarters, parlor, ca. 1890s. The piano would have been a prized possession. Most of the rest of the furniture is light and portable, and decorative effects are achieved with cloths and carpets. Courtesy Library of Congress.

looking bedstead in the bedroom. A pine table in the dining-room and a range in the kitchen completed the outfit. A soldier had scrubbed the rough floors with a straw broom: it was absolutely forlorn, and my heart sank within me." From a chaplain who was leaving the post she bought "a carpet . . . a few more camp chairs of various designs, and a cheerful-looking table cover." Similarly, when Elizabeth Burt reached Fort Bridger in 1866 and moved into the log quarters of the commanding officer, she created her own domestic surroundings. "After the cleaning was finished and our household effects in place, there was quite an air of home comfort about the house. A Brussels rug was on the floor of the living room, while four army blankets sewed together made rugs for the other rooms. Bedsteads were found in the house with a few tables and chairs. Curtains were made of pretty material brought with us, which also covered packing boxes to serve as toilet tables and washstands." Frances Carrington passed through Fort Laramie briefly in 1866 and with her husband was assigned quarters in an adobe building. As in Elizabeth Burt's house, "gray army blankets

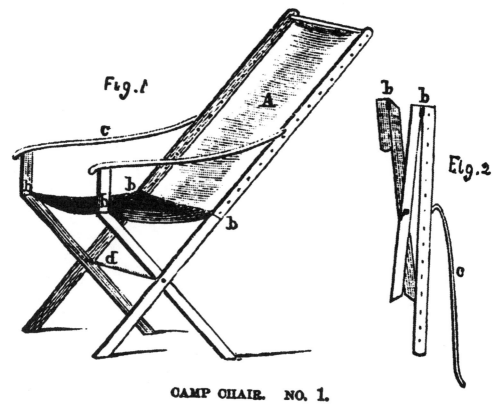

Fig. 5-3. Camp chair, 1859. Veterans advised that officers furnish their quarters with folding chairs such as these. From Marcy, *The Prairie Traveler*, 145.

were tacked upon the floors to the extent of their capacity. Hospital cots were used for beds, and we began, as the children say, 'to keep house.'" She took the opportunity to unpack: "And now for my nick-nacks and such belongings to reproduce home environment, as I attempted once before on Governor's Island, where my initial army experience began!"[5] In the act of recounting these domesticating gestures, these officers' wives were validating their own existences, as exercising their civilizing influences was their primary reason for being there.

Neither Martha Summerhayes nor Elizabeth Burt did all the work themselves; both had servants. Burt brought her own servant, a former slave, with her; the servant lived with the family. Summerhayes had the assistance of a soldier assigned to the task. "Strikers," as they were called, received additional pay from an officer to work as his servant; this practice was outlawed in 1870, but obviously not enforced. In 1881 General Sherman argued that officers deserved servants, and because officers were often posted in places where servants were hard to obtain, even to the extent that servants' wages exceeded officers' pay, officers should be able to hire servants from among the soldiers, providing that the soldiers and their commanding officers agreed. Summerhayes's striker, a man named Adams, did not know how to cook any better than she did. Summerhayes went to the quartermaster storehouse to select kitchen equipment, where she found "nothing smaller than two-gallon tea-kettles, meat-forks a yard long, and mess-kettles deep enough to cook rations for fifty men!" She struggled with the equipment, she struggled with cooking at the high altitude, and she struggled with her own inadequate education that was "more engrossed by the study of German auxiliary verbs, during the few previous years, than with the art of cooking." Finally, she took the advice of other wives and bought the appropriate pots and pans in Cheyenne, and her cooking gradually improved.[6] With efforts such as these, officers' wives created the civilized surroundings that were expected of them. They used their furnishings to set themselves apart from enlisted men and their wives, from Indians, of course, and also from some of their neighboring settlers.

Although at times post quartermasters covered the walls of officers' quarters with boards rather than lath and plaster, in general the interior finish was paint or wallpaper. In 1860 the deputy quartermaster general disallowed the request of Maj. Edward R. S. Canby, commanding officer of Fort Bridger, for 100 pounds of chrome yellow, recommending instead "a mixture of pulverized charcoal, with the white-wash . . . giving a dark slate color, not unpleasant to the eyes." But by the 1880s, colors ran riot. At Fort Bridger in 1884, when officers' quarters, barracks, and the commanding officer's quarters were under construction, the post quartermaster ordered the following paints: white lead, burnt umber, raw umber, burnt siena, raw siena, copal varnish, boiled oil, turpentine, oak graining color, walnut graining color, and whiting. Soon thereafter, another post quartermaster ordered vermilion, scarlet lake, burnt siena, and prussian blue paints, and oak and black walnut graining colors. A few years later, the post quartermaster requested wallpaper samples from two firms in Omaha. He noted that the rooms he was papering were small, only 7–8 feet in height, so he wanted light shades and small figures. Going with the lowest bidder, as contracting procedures required post quartermasters to do, could be a problem. Post Quartermaster Eltonhead refused to buy wallpaper from the lowest bidder, preferring another design for the four rooms occupied by two captains.[7] Although the effect of this decorating is unknown, as the interiors of quarters were rarely photographed or described, the orders sounded colorful, contemporary, and decidedly nonmilitary.

The well-documented furnishings of enlisted men's barracks, by contrast, caused complaint and, eventually, regulation (fig. 5-4). Until after the Civil War, soldiers slept in double bunks—two men to a bunk, and two tiers high, sometimes three. The "bunkies" slept head to foot, sharing blankets.

Fig. 5-4. Double bunks, 1864. This drawing of a "stag dance" in a log barracks shows wooden double bunks in two tiers. From *Harper's Weekly*, 6 February 1864.

This proximity to fellow soldiers repulsed at least one officer, who wrote to the *Army and Navy Journal:* "Now, I must not be accused of prudery or squeamishness, when I say that this style of packing the men is decidedly an improper one." Similarly, one soldier asked that "provision be made for the men to sleep singly and alone and not keep up the present barbarous and unhealthy system of having the men sleep in couples summer and winter." There were other objections, other than overfamiliarity, to double bunks: bed bugs infested the wooden bunk and straw bedding. Inadequate ventilation was the concern of John S. Billings of the Surgeon General's Office. Calling double bunks "evil" and "relics of barbarism," he noted that they were in use in ninety-three, or more than half, of the posts in 1870. "It is certainly time that the use of such bunks should be absolutely and imperatively forbidden, and so long as they are allowed to exist in dormitories, so long it is useless to hope that those rooms can be made what they should be. No one acquainted with the first principles of sanitary science will approve of their use. They have long been discontinued in the service of European armies."[8] Gradually, single iron cots replaced the double wooden bunks.

The shift from double, two-tiered bunks to single bedsteads was facilitated by the downsizing of the army in the post–Civil War period, which meant that fewer men occupied each barracks dormitory. The quartermaster general introduced iron bedsteads for hospitals in 1858. By the 1870s the army made single cots available for barracks. In 1871 the quartermaster general bought iron bedsteads from three suppliers and in 1872 purchased 8,666 iron bunks from Composite Iron Works in New York City. By 1873 the changeover was complete, and nearly all posts had been supplied with iron bedsteads. As the quartermaster general noted, "They give each soldier a separate and distinct bed, and conduce both to comfort and health, and are a great improvement upon the rough wooden two-story bunks heretofore in general use at military posts." There were signs, however, that the single bedstead was not universally welcomed by the men. The post surgeon at Fort Bridger observed that "to permit the men to club their blankets during the protracted winter," the beds were pushed together in pairs.[9]

The 1872 model Composite bunk was a single, one-level bedstead, with iron headboard and footboard; four wooden slats, 6 inches wide and 6 feet 10 inches long, supported the bedding. In 1878 the Quartermaster General's Office adopted specifications for the Coyle bunk, which was made of galvanized iron pipe, including a side rail. Wooden headboard and footboard, as well as the four wooden slats specified earlier (although now only 6 feet 7 inches long), completed the simple bed. The Coyle bunk weighed only 32 pounds, compared to the Composite's 61 pounds, offering an obvious advantage. In 1885 Captain John F. Rodgers, the army's military storekeeper, mentioned what he called "quite a radical, and it is hopeful, beneficial, change": the introduction of wire-woven bunk bottoms, to replace the wooden slats. In addition, mattresses and pillows "of good strong ticking filled with cotton linters" replaced the self-made straw mattresses. The army provided two covers for each mattress, as well as sheets and pillowcases.[10]

Before the introduction of the iron bedstead, soldiers made their own wooden bunks in accordance with a common design (fig. 5-5). Other barracks furnishings were less standard. The men used company funds—from a surplus created by pooling the men's flour allotment—to purchase amenities and supplies for other constructions in the barracks, as well as for additional food. Besides extensive equipment for eating and cooking, one company used its fund to buy a clock, a metronome, newspapers, and 90 yards of calico. Contemporary photographs show curtains of brightly patterned cloth covering shelves, not windows (fig. 5-6). Capt. John G. Bourke described barracks at Fort Grant, Arizona, as being decorated with Apache weapons, Navajo rugs, newspaper illustrations, and a pet mockingbird in a wicker cage. Soldiers used lumber from packing crates to make furniture,

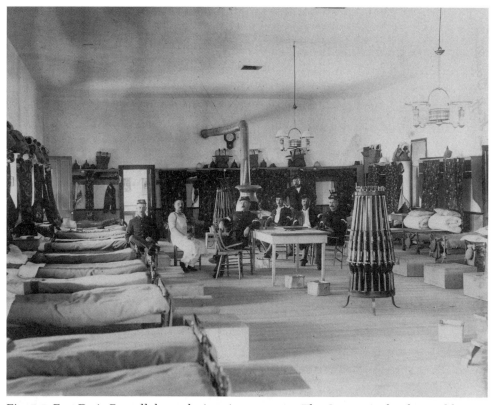

Fig. 5-5. Fort D. A. Russell, barracks interior, ca. 1895. The Composite bunks, coal heater, barrack chairs, oil lamps, and footlockers are all standard issue. Brightly colored curtains over shelves, the clock, and spittoons are not. Courtesy National Archives.

and then tried to take the chairs, tables, stools, and shelves with them when they moved.[11]

Finally, in 1878 the army introduced barrack chairs as standard issue, providing one for each noncommissioned officer above corporal, and six for every twelve other enlisted men. Fabricated at Fort Leavenworth, the chair was "a strong, substantial wooden chair" with a molded wooden seat. "It is easy, durable, and cheap, and will add much to the comfort of troops." The quartermaster general also noted that soldiers "have heretofore been accustomed to sit on benches or boxes or their beds." The chairs received hard use; in two years the quartermaster general distributed iron bolts and braces to reinforce them, as they "in many cases became rickety in the joints, and in some broke down entirely." In an effort to make the chair lighter, in 1883 a new model was issued with a rawhide seat. However, durability continued to be a concern, for the new version also had iron rods.[12]

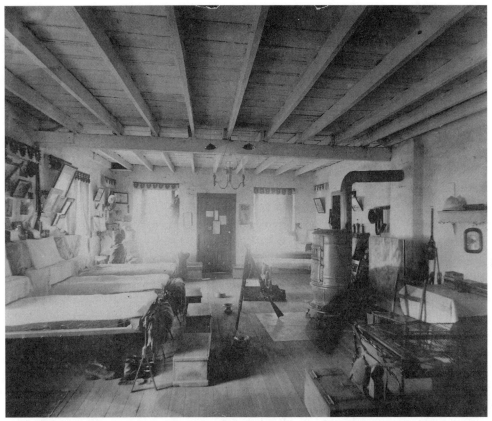

Fig. 5-6. Fort Shaw, Montana, barracks interior, photo by C. Eugene LeMunyon, ca. 1888–1901. Composite bunks with wire-woven bottoms are evident in this barracks of an African American company, whose members decorated the walls with photographs and prints. Courtesy Montana Historical Society.

Heating and Ventilation

In the view of army surgeons, ventilation was a critical aspect of modern medical practice. It was also a point of some contention. The soldiers thought they suffered from too much ventilation, not too little, while the Surgeon General's Office advised more ventilation. Eliminating "poisons" in the air required both adequate air space and a means of changing the air. Army regulations did not fix air space, but did determine floor space. Based on a ceiling height of 10 feet, the army called for 375 cubic feet in the north, and 425 cubic feet in the south, per soldier. Billings, after surveying the requirements of other nations' militaries, recommended 600 cubic feet in the north and 800 cubic feet in the south. To change the air, Billings relied on a sufficient number of windows; he recommended one window for every two beds in a dormitory 24 feet wide. In the winter, heat aided ventilation; air going up the chimney was replaced by fresh air, which had to be warmed, but a flow of air was established. As Billings advised, "To secure the greatest effect from heat as a ventilating power, the fresh, cool air should enter at the bottom of the room and the warm, foul air pass out above."[13] To achieve this, Billings recommended the ventilating double fireplace for barracks, as he had for hospitals.

As far as controlling the system of ventilation in barracks, the Surgeon General's Office was limited to an advisory role. With no right to approve barracks designs, Billings and the post surgeons could only complain about what had been built. Billings criticized Meigs's model plans issued in 1872: "This plan gives about 500 cubic feet per man, has no arrangements for ventilation, and no provision for bath-rooms." He went on with a backhanded compliment and a veiled threat: "While this is a great improvement on the majority of existing barracks, it cannot be considered perfectly satisfactory nor will any similar attempts at economy by reducing the air-space in the dormitories prove a success." Just such an "attempt at economy" was made in the construction of the two-company concrete barracks at Fort Laramie, based on the model plans. The post surgeon recognized variations from the model plans: "These are constructed on the general plan recently ordered for barracks, except that the first story is only 10 feet and the second story 9 feet high. This is a very serious error, as it reduces the air space in the dormitories, makes them look low, and not symmetrically proportioned. The ventilation also is defective, there being only two very small shaft-ventilators in each dormitory."[14] The model plans indicated a 10-foot, floor-to-ceiling height on both floors and no ventilating shafts.

In spite of the post surgeons' misgivings, the men who lived in these barracks felt the space was amply ventilated. As the post surgeon described adobe and wood barracks at Fort Laramie: "The cracks in the floors are wide, and in winter the air which comes up through them makes the rooms not a little uncomfortable. . . . A few of the men have buffalo robes. The most of them are fain to protect themselves against the rigor of the winter by eking out their scanty covering with their overcoats. They nearly all complain of sleeping cold." At Fort D. A. Russell, the post surgeon noted that natural ventilation blew through the adobe-lined, wood-frame barracks: "The roof and ends above the adobe lining are so open from the shrinking of the lumber that ventilation is amply sufficient. Some of the barracks have, however, special ventilation shafts."[15]

The men often subverted ventilation systems. The post surgeon at Fort Fetterman noted that there was "no special arrangement for ventilation except in one barrack, where four holes two inches square are arranged over each window with slides, but they have been carefully nailed down, and canvas tacked over them, the men complaining of draught." Charles Smart at Fort Bridger noted that "during four months of the year the temperature falls below zero at night. The great object of men in quarters in such weather is to keep warm. The fires are well attended to, and every chink which admits fresh air, and can be reached, is carefully stopped up."[16] Unsympathetic to the cold that the men felt, the doctor still insisted on ventilation.

The belief in the health value of ventilation also promoted the use of large barrack dormitories rather than the dividing up of the company among smaller rooms. Although older barracks, such as those at Fort Laramie built in 1850, often had several smaller sleeping rooms, after the Civil War large dormitories became universal. Lt. Col. Thomas M. Anderson found widespread support for smaller rooms among both officers and enlisted men. He envisioned rooms that would sleep twelve to sixteen men, who would be able to choose their companions, thus separating out the good from the bad and promoting contentment among the troops. But the belief that only large rooms could be well ventilated, as well as the greater cost of heating more rooms, militated against the adoption of this system.[17]

Ventilation in winter depended on the heating system, which consumed oxygen and assured air flow. At both Fort Laramie and Fort Bridger, brick fireplaces heated all of the original residential buildings. After the Civil War, the army began constructing brick chimneys for stoves, such as the barracks and quarters at Fort D. A. Russell built in 1867–1868, which had provision for stoves and no fireplaces. The army also retrofitted older buildings with stoves, blocking up the old fireplaces, as at Fort Bridger, where wood-fired stoves replaced fireplaces by 1870. In 1873 coal-fired stoves replaced wood-fired ones. Although stoves consumed less fuel than fireplaces, Assistant Surgeon John Shaw Billings objected to coal stoves, especially "their contaminating the air with carbonic oxide gas," or carbon monoxide (CO). He also criticized the air flow, preferring the greater drafts of an open fireplace. Billings's desire to increase ventilation stood in direct opposition to the goal of providing a more efficient heating system.[18]

Billings lost this argument. The army continued to supply stoves to its posts. In 1876 the quartermaster general issued the first specifications for stoves. Standard barracks adhering to the 1872 model plans were allocated wood-burning stoves as follows: two large stoves in the dormitory on the second floor; one large stove in the mess room; one large stove in the day room; small stoves in the first sergeant's room, company orderly room, and saddler's shop; and a cooking range in the kitchen. Regulations provided officers with stoves, according to the number of rooms they received by rank, beginning in 1881. Thus, a lieutenant was allocated one stove for heating and one for cooking, to accommodate his one room and a kitchen. Fuel allowances had been granted beginning with the 1861 regulations, allowing a lieutenant one-half cord of wood per month in the summer and two in the winter, with increases for higher latitudes; the equivalent in coal was also specified. By 1889, however, officers were required to pay for their allotment at a fixed price; any fuel over that allotment could be bought at the contract price.[19]

Sanitation

Sanitation improvements separate modern Americans from their nineteenth-century counterparts. Today's standards of cleanliness differ greatly from those in the past, and modern Americans tend to associate good sanitation with modernity. Fort conditions produced the same sort of muddling progress to cleanliness and running water that they did toward modern architectural practice. This was partly because sanitation was not a high priority and also because technological expertise was lacking. The uncertain progress of the forts placed them in between their rural and urban neighbors in the adoption of sanitary measures.

Sanitation received less attention from the Surgeon General's Office than ventilation, although post surgeons reported on post sanitation regularly. Post surgeons not only directed all the operations of the hospital, but also inspected the entire fort for sources of sickness and poor sanitation, and made recommendations to the commanding officer of the post. Human waste was a frequent concern of post surgeons. The most common method of disposal was a latrine, often called a "sink" in army terminology, which consisted of a hole in the ground, over which stood a wooden building that provided seats. Earth closets were popular for a short time after the Civil War; these provided earth to cover excretions and absorb odors. Water closets, in which water flushed excretions into a cesspool or into a sewer system, required running water.[20]

The simplest method was the latrine. As the post surgeon at Fort Laramie described it: "Each set of barracks is provided with a sink. These sinks are wells with rough board houses built over them. The houses [are] furnished with seats and urinals. They are kept in tolerably good order and lime is thrown into them once a week." A similar system prevailed at Fort Bridger: "The sinks are built in rear of the quarters. Vaults are used, and as they fill near the surface, are covered and new ones dug." There was a limit to how many times a latrine could be moved, however. An inspection of Fort D. A. Russell revealed the following: "Both barracks and quarters now depend on vaults, and the ground in rear of each is honeycombed with abandoned and filled up pits."[21]

The shortcomings of this system are revealed by attempts to change it. The Surgeon General's Office urged the adoption of the "dry-earth system," or earth closet. At Fort Laramie in 1885, Post Surgeon D. G. Caldwell recommended a multi-company general sink located near the river. The fire engine could flush it, and it would drain into the Laramie River. A year and a half later, Post Surgeon Louis Brechemin rejoiced at its construction: "The completion of a new privy for the use of the Band, Guardhouse and Companies A, F, H, & K may be mentioned as a marked sanitary improvement. As the vault [is] constructed of concrete with the necessary pipes for the daily flushing of the contents into the adjoining river, danger of soil pollution is prevented. A great advantage over the old system of sinks for many years in use at this post is gained in this way."[22] As great an improvement as this general sink might have been, the modern convenience of the time was the water closet. This had to wait until there was a steady water supply and some means of waste disposal.

In 1870 the post surgeons expressed great satisfaction with the quality of the water at their respective posts, as well as with the method of distribution, which consisted of wagons delivering river water to barracks, quarters, kitchens, and laundries. As Fort Laramie's post surgeon explained: "The water used for culinary and household purposes in the garrison is chiefly obtained from the Laramie River above the post, and is hauled around in a large tank on wheels and dispensed as necessity may require" (fig. 5-7). Fort Bridger used "a few surface ditches along the regular slope, by which [water] is made to run convenient to all the barracks," except in winter, when soldiers used a wagon. Water delivery was not a pleasant duty; the task usually fell to prisoners. Elizabeth Burt observed during a bitterly cold winter: "My deep sympathy was appealed to each day when the men brought the water from the river in the wagon sheathed in ice. Of course the water froze as it spilled on the men." The water was as difficult to obtain as it was to dispense. The post commander argued for "a proper system of water supply": "When you reflect that during the winter months water is only obtained by cutting through the ice—which is reformed in a few hours—you will readily realize the hardship of watering the animals and furnishing the command." At Fort D. A. Russell, a pump raised water to a tank house, from which it was distributed by wagon: "Attached to a saw-mill situated in the bottom [of the Crow Creek streambed], west of the post, is a steam-pump which forces the water from the stream up into an elevated wooden tank on the bluff. A water-wagon is filled daily from this tank, and delivers the water for the officers, enlisted men, and laundresses into barrels near their quarters. The water is also made to flow from this tank through a ditch around the parade, thus supplying the trees there planted." Officers and men preferred this distribution system to wells: "There are wells behind the officers' quarters and barracks, but they are not used, because they run dry in summer, and the other plan of supply is at all times more convenient."[23]

The barrels filled by the water wagon provided water for washing people and for washing dishes. Thus barracks sometimes had "washrooms" with basins for the men to scrub their faces and hands; if not, this washing took place in the dormitory. Bathing was performed much less often. Unless men could bathe in streams, they used wagon water to fill tubs in the washroom or other available space. In officers' quarters servants carried basins and pitchers to family bedrooms for daily washing. Similarly, servants filled tubs placed in bedrooms when the family desired baths. The kitchens of both barracks and quarters required water for cooking and cleaning. Laundresses likewise needed water for personal use, as well as for their work. They often established themselves near a river for easy access to water with which to fill their tubs. Livestock also required a regular water supply.

Running water was a convenience common to the urban middle class in the 1880s, and army officers in the West thought they deserved no less. The first effort at improvement, though, was directed at finding a steady water source for Cheyenne. In an apparent pork-barrel project encouraged by the governor of the territory of Wyoming, Congress appropriated $10,000 for an artesian well at Fort D. A. Russell in 1872. As Lt. Gen. Philip Sheridan noted, the "results if successful will be of great advantage to the people of Cheyenne." The secretary of war gamely termed it "a good experiment for an arid country." Unfortunately, the experiment failed. Nine months after drilling commenced on November 30, 1872, the well was at a depth of 600 feet and still had not hit water; the post quartermaster noted that it was "the only well known to have reached the depth of 600 feet without finding rock." By April 1874, when the secretary of war called off the experiment, the well had reached 1,420 feet without finding water.[24] With the failure of the artesian well experiment, both the army and the civilians of Cheyenne continued to take their water out of Crow Creek.

Water rights have always been important in the American West. Most army posts were set in remote locations near a river and did not have to compete with anyone for the use of that water. But a sizable city grew up near Fort D. A. Russell and competed for its water supply. The fort used water in two ways: for household use, for which water pumped up from the creek to a tank was distributed by wagon, and for irrigation in the summer. By the 1880s, a dam 2 miles up the creek diverted water into a

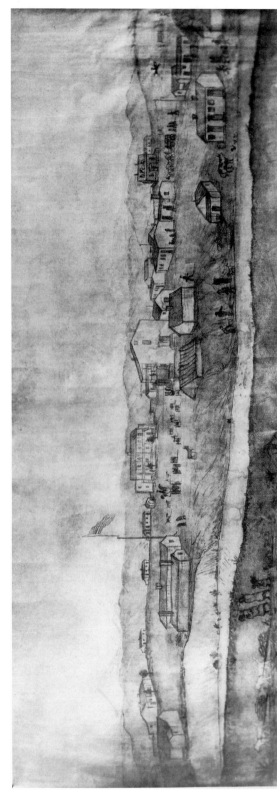

Fig. 5-7. Fort Laramie, drawing by Caspar Collins, 1864. In front of the post trader's house, the dark-colored building on the right, is the water wagon, described by Catharine Wever Collins in a letter to her daughter Josie dated January 16, 1864: "The water wagon with three white mules is followed by three prisoners in charge of the guard who carries his gun under his arm as the thermometer is 20° below zero, and has his hands in his breeches." Spring, ed., "An Army Wife Comes West," 16–17. Drawing from American Heritage Center, University of Wyoming.

ditch for irrigation purposes. The city also drew from this impoundment, and soon prevented the post from receiving any irrigating water. In 1884 city officials and the fort's commanding officer agreed that the city would provide water to the post in exchange for a ditch and right-of-way across the military reservation, so that the city could continue to take water from the creek above the fort.[25]

In 1885 Fort D. A. Russell received a water system that included a 4-inch pipe running around the post, providing water to hydrants behind the barracks and quarters (fig. 5-8). Just after the first group of six brick captains' quarters was constructed in 1885, James Regan, the post quartermaster, proposed that additional captains' quarters be built with "bath houses," and that the same be added to the quarters already constructed, at a cost of $298.47 each.[26] The "bath house" he proposed was a room on the first story with a bathtub but no sink or water closet. Presumably, the tub would be filled by water hand-carried from the outdoor hydrant. This plan was not adopted, and the four additional captains' quarters were built in 1888 without bathing facilities.

Fort Bridger obtained its water system with seeming reluctance. In 1883 Lieutenant Colonel Anderson, the commanding officer, reported that he had received a wooden water tank of 18,000-gallon capacity and a small pumping engine, which were to be used to fill the water wagon (fig. 5-9). He foresaw a system of piping, but did not try to build it at that time. In May 1885, post quartermaster Charles Stivers requested approval for a waterworks and piping. The quartermaster general approved the request in August, but no action could be taken before winter set in. By the next summer, both the post commander and the post quartermaster had changed, and the foot-dragging began. In August 1886 the new post quartermaster, Francis Eltonhead, explained that the laying of the water pipes had not begun, and that he did not want to begin it because the post lacked a plumber: "I am unwilling to undertake the laying underground of some $2,300 worth of pipes and then find the water-system of the post a failure, and no one has so far undertaken this work, although all the materials have been here for some time past." Prodded by the chief quartermaster, by November he started laying pipes and experimented as to the depth at which pipes had to be laid to be below the frost line. By July 1887 Chief Quartermaster Dandy demanded an explanation: "It is understood that post is still supplied with water by means of a water-wagon, although the required material for a system of water works was purchased and shipped to that post in 1885." That fall, Eltonhead toured the water works of Rock Springs, guided by an engineer of the Union Pacific Railroad, and learned how to design a water pipe to cross a stream. In October a new chief quartermaster, William B. Hughes, could not understand the delay: "Telegraph

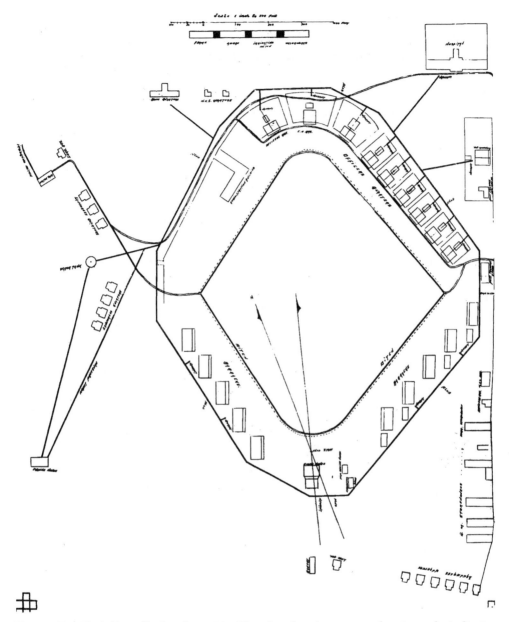

Fig. 5-8. Fort D. A. Russell, site plan, 1884. The plan showing proposed waterworks indicates two distribution systems, one in ditches for irrigation and one pumped up to the water tank and piped to hydrants behind houses and barracks. Courtesy F. E. Warren Air Force Base.

when you are ready for the services of an expert plumber to lay water pipe. What are you waiting for?"[27]

Finally, in May 1888, help arrived in the form of 1st Lt. Hiram M. Chittenden of the Corps of Engineers and plumber S. J. Murphy. All available soldiers dug ditches in which to lay pipe, a job the men found disagreeable:

"to dig in ditches almost knee deep in water is far from pleasant." When Chittenden departed at the end of June, supervision fell to 1st Lt. Alexander Ogle, post quartermaster. Murphy stayed until mid-July, and oversaw the raising of the tank house to increase the water pressure. The water works was completed by early August. One of the drawbacks of the system was that soon after its completion, the post quartermaster requested a "competent civilian engineer" to run the engine; the enlisted man who had that job had deserted. Ironically, the post quartermaster received a request from his counterpart at Fort Duchesne, Utah, who wanted the plans and specifications of Fort Bridger's water system, so that he could copy it for a similar system at his post.[28] With all of the problems and inexperience associated with the construction of the water system at Fort Bridger, the post's apparatus was still more sophisticated than others.

Fort Bridger's five-year saga to build a system of piped water reveals several aspects of military construction. For one, expertise was clearly lacking; none of the several post quartermasters felt comfortable supervising such a complex undertaking. Second, the personnel changed frequently: the post commander, post quartermaster, and chief quartermaster in Omaha all changed during the five-year period, disrupting continuity. Such frequent job changes may have contributed to a willingness to delay. There may also have been legitimate concerns, such as the high water table and extreme temperatures found at Fort Bridger.[29] Despite all the problems, it is surprising that something as convenient as piped water would not have been more of a priority for personnel at the post.

The provision of piped water was also important for fire suppression. In 1883 Fort Laramie acquired a pump and boiler, along with pipe and fittings that had been used at the Yorktown Centennial. A steam engine pumped water from a well below river level up to a tank, from which water was distributed by wagon. In 1885 Fort Laramie's post quartermaster successfully argued for a special requisition to obtain pipe and hydrants "for the purpose of supplying water in case of fire, and for irrigation." The six fire hydrants installed that year were presumably also used to provide water for drinking, cooking, and washing.[30]

With a steady water supply came the problem of water and sewage disposal. Cesspools, in which sewage was deposited in the ground, were often viewed with alarm because of the perceived miasmatic effects of poisons in the earth. A letter to the *Army and Navy Journal* cited an article in the *Sanitary Engineer,* which held that sewage in open gutters was preferable to a cesspool, which "would pollute all the ground all about it with putrescent organic matter, which sooner or later would be likely to cause harm to the inmates of the house, reaching them either through the drains or the water supply or both." In a gutter "the effluvia is soon diluted in the air, or

Fig. 5-9. Fort Bridger, 1889. Visible in the far background are the water tank and engine house, puffing smoke. Courtesy Wyoming State Archives, Department of State Parks and Cultural Resources.

washed away by the rain." Such may have been the concern of the acting inspector general who examined Fort Laramie and reported that "in the commanding officer's quarters the pipes from the bath tub, standing wash stand and water closet lead to a cess pool in close proximity to a cistern and are without traps."[31]

At Fort D. A. Russell, post quartermaster Edward Chynoweth tried to gain approval for a sewer system in 1888, submitting an estimate for $11,718.25. The chief quartermaster rejected his proposal as inadequate and suggested that the problem be assigned to the depot quartermaster, Charles F. Humphrey. Humphrey devised a plan that involved connecting with the city's sewer system at the corner of 21st and Eddy Sts. The following year, the quartermaster general approved $25,000 for its construction. Again, the civilian interest in this project was seen in a query from Wyoming congressional delegate Joseph M. Carey, wondering why the contract had not yet been let.[32]

The completion of the sewer system in 1890 was followed closely by a requisition for 333,000 sheets of toilet paper. As the post commander, Col. Henry Mizner, explained, "The use of proper toilet paper is deemed essential to guard against the choking and damage of the new Sewerage by the use of newspapers." Apparently the chief quartermaster had never received such a request, because he forwarded it to Washington, noting that "while water closet paper does not appear to be a proper charge against the Quartermaster's Department," he still thought it necessary. In Washington, the assistant quartermaster who received the request recommended that "the

Depot Quartermaster New York be instructed to obtain and submit prices for the various forms of W. C. Paper purchased in quantity, with a view of placing a stock in Depot from which to issue this."[33] Evidently, Fort D. A. Russell was breaking new ground in its request for toilet paper.

Once a post had running water and an adequate sewage system, installation of bathrooms in the officers' quarters was the next priority. Water closets required a new room, as there had not been an indoor space allocated to their predecessors. The bathroom usually included a water closet, bathtub, and basin. The boiler next to the kitchen stove heated water. After several proposals had been rejected, the new brick captains' quarters, constructed in 1888, received bathrooms in 1891. Rather than put the bathroom in an addition on the first floor, which was one of the proposals, the bathroom was placed in the half-story above the kitchen; a large wall dormer was added to accommodate it.

The delay in building a sewer system at Fort D. A. Russell, which slowed the installation of plumbing in officers' quarters, is somewhat puzzling. There is no doubt that officers wanted sanitary conveniences; at Fort Douglas, Utah, in 1890 a lieutenant bought and installed his own water closet. When he was transferred, his superiors ruled that it was a permanent improvement, and he could not take it with him, although he had paid $35 for it. Lack of expertise at widely scattered posts may have been part of the problem. In 1888 field officers' quarters constructed at Fort D. A. Russell were equipped with full bathrooms, but not until someone with sufficient expertise was brought in to create new drawings and specifications. A staff officer argued that Quartermaster General Holabird was doing much to improve water supply and the education of his staff, but this observer recommended that a special "division of architecture, sewerage and water supply" be created within the Quartermaster General's Office, recognizing the new professionalism that these new technologies required.[34]

While the army apparently distributed water closets according to rank, the field officers receiving them three years before the captains, conditions lagged even further for noncommissioned officers and enlisted men. At Fort D. A. Russell the noncommissioned officers' quarters built in 1885 soon needed bathroom additions. In 1890 the quartermaster general ruled that plumbing in noncommissioned officers' quarters should be restricted to the bathroom and water closet; hot water circulation was not necessary, as the small amount of hot water required could be obtained from tanks on the backs of kitchen stoves.[35] He did not explain why noncommissioned officers' families would need any less hot water than officers' families. Fully equipped bathrooms were even less of a priority for enlisted men. The recycled frame wings in the rear of the new brick barracks built in 1885 were remodeled to include a washroom, equipped with basins and a bathroom with two tubs. Enlisted men continued to use latrines until their barracks received new brick wings in 1896.

In the 1880s army officials were more concerned with providing bathing facilities to enlisted men than flush toilets. Bathing was not a concern of the frontier army in its earlier years. Fort Laramie's post surgeon explained the situation in 1870: "The men bathe freely and constantly, in pleasant weather when off duty, in the stream above the post. There are many places in the river where the water is ten to twelve feet deep, affording opportunities for swimming. No bath-houses have as yet been erected, principally because their need has not been felt." The post surgeon did not explain why no need was felt in the winter. That same year, Assistant Surgeon John S. Billings deplored the absence of bathing facilities. He asserted that "at least once a week every man should thoroughly cleanse his entire person," and urged that bathing facilities be provided at every post. Through the 1870s, however, the secretary of war routinely rejected requests for bathing facilities.[36]

In 1875 Billings criticized Meigs's model plans for failing to provide bathing facilities for either officers or enlisted men. He claimed that "a dirty man will, in most cases, be a discontented, disagreeable, and dissolute man; for the condition of his skin has much more to do with a man's morals than is generally supposed." He advocated the provision of bathtubs for winter use, though he felt swimming was preferable in the summer. He recognized the problems of providing bathhouses as being "deficient water-supply" and "difficulty in heating the room." To solve this problem, he recommended the use of "jets or showers," and sketched an eight-stall shower with a stove in the center for heating the room and the water. Billings was ahead of his time, however, and showers did not become standard at army posts until after the turn of the century.[37]

Bathrooms, to which water was delivered by wagon or carried by hand, were usually located in buildings adapted for the purpose. At 1883 at Fort Laramie, each company had a building that combined the bath, reading, and washroom functions. The combination of bath and reading rooms in one building perhaps signified a philosophy of "clean mind in a clean body," or just a conjunction of personal benefits.[38] In 1879 Post Quartermaster James F. Simpson had proposed a twenty-tub bathhouse with "French cast iron bath tubs" for Fort Laramie, but it was not approved. As regimental quartermaster for the 3rd Cavalry, Simpson accompanied his commanding officer, Col. Albert G. Brackett, when he was posted to Fort D. A. Russell the next year. There, he submitted the same drawings for a bathhouse, where again it was disapproved (fig. 5-10). Finally, in 1888, Depot Quartermaster Charles F. Humphrey oversaw construction of the bathhouse at Fort D. A. Russell, which the acting inspector general described: "A fine bath house has recently been erected for the command, but it is inconveniently situated, being near the pump house." The simple frame building had eighteen tubs, with two tubs assigned to each company, one to the band and one to the guard house.[39]

With the advent of running water and the dismissal of laundresses, post quartermasters began to propose construction of steam laundries. Post Quartermaster James Regan of Fort D. A. Russell explained that steam laundries were "very desirable, especially now since the great improvements in the men's bedding, etc. Washerwomen are few at the post and the ones now remaining are not always reliable." Fort Laramie's post quartermaster saw steam laundries as a way to rid the post of the women who no longer functioned as laundresses on the army's rolls. He explained, "This would take away the chief source of income of the women who now hang around military posts and thus deprive the soldiers of the principal inducement to marry.... I think steam laundries would do more to abolish them than any orders." He also proposed a way of providing steam laundries: "At this post the necessary power could be furnished by the boiler and engine used to saw wood.... Bath rooms should be erected in connection with the laundries which could be done at very small expense."[40] In his view, the boiler at the sawmill would provide steam to the laundry, and heat water for bathing as well.

Lighting

Lighting was yet another instance of the army lagging behind urban civilians. Interior illumination at army posts was by candlelight until 1881, when the army first issued lamps. Oil lamps were not unknown earlier; Elizabeth Burt brought some oil lamps to Fort Bridger in 1866, but could not find a supply of oil there. "The Quartermaster had no coal oil and it was too costly at the post traders to be used more than very sparingly. We bought candles from the Commissary." Candles were issued to enlisted men at a rate of twenty ounces to one hundred rations, or about three candles daily to a company. The quartermaster general issued a "Report on Lighting Company Quarters" in 1879, which observed that "it is reasonable to expect that the soldiers will find their quarters dark, cheerless, and uninviting." Part of the problem was that stoves had replaced fireplaces, which had cast more light; a stove "gives not that blaze of light and cheer they were wont to enjoy in former days from big wood fires in spacious fireplaces, that lighted and warmed their rooms."[41]

In 1881 lighting underwent a major change. Lighting responsibility shifted from the Subsistence Department to the Quartermaster Department,

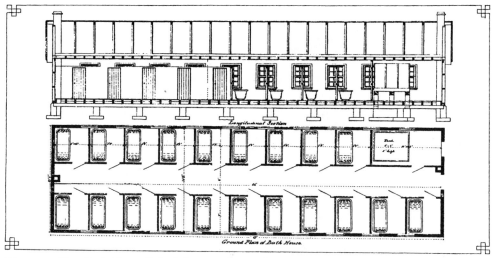

Fig. 5-10. Fort D. A. Russell, bath house plan and section, 1880. Although the construction of this bath house was disapproved, a similar bath house was built eight years later. Courtesy National Archives.

which replaced candles with oil lamps throughout the service. The secretary of war claimed that the oil lamp gave the light of sixteen candles, thus justifying the change on the grounds of efficiency. After testing three lamp suppliers, the Quartermaster Department selected the Manhattan Brass Company in New York, which manufactured lamps at a cost of $4.40 each. The Quartermaster General's Office developed specifications for two styles of lamps: a two-light pendant and a single-light bracket lamp. Fueled by mineral oil, or kerosene, the lamps were supplied to the posts at a rate of one burner to every ten men, with additional lamps to noncommissioned officers, post libraries, and post schools. Within a year, the Quartermaster Department supplied 5,156 bracket lamps, 1,782 pendant lamps, and 2,004 lanterns to posts.[42]

The decision to change the means of lighting was based not only on efficiency, but on a moral argument. The secretary of war argued that "the *morale* of the troops is reported to have improved at posts where the new lights have been introduced. The men, being able to read without injury to their eyes, spend more time in rational amusements and less time at the sutler store, at the grog-shops, and in the guard-house." The enlisted men apparently agreed, if this poem written by a soldier is any gauge:

> So if "fiat lux" the order is,
> And candles are shown the door,
> Round the bright kerosene twenty men will be seen,
> To one at the trader's store.

In 1881 the *Army and Navy Journal* reported that "the officers of the [quartermaster] department feel confident that the greater facilities for a profitable employment of the evening leisure of the soldier, afforded by the introduction of these lamps, will elevate the moral tone of the men, keeping them out of low resorts, making them contented and happy, and promoting military discipline." The dubious argument that better lighting would keep soldiers reading in their barracks rather than going to bars was apparently a compelling one. But it was a different argument from that put forth thirteen years later by the inspector general, that "barracks should be bright, cheerful, and inviting," which implied that comfort alone was enough to justify improvement.[43]

Within a few years of the introduction of oil lamps, complaints abounded that there were not enough of them, and that they were not casting enough light. Gas lighting was rare at western posts, as it required a gas plant and extensive piping. As a result, forts retained oil lamps, for both interior and exterior lighting, long after most cities provided gas lighting. In 1893 the inspector general of the Department of the Missouri argued that "each post having a steam-heating plant should be lighted with electricity." He mentioned in particular the well-established forts of Sheridan, Riley, and Leavenworth.[44] Fort D. A. Russell received electricity in 1904, nearly twenty years after the city of Cheyenne was electrified.

Novelist Owen Wister, who visited Cheyenne in 1885, lamented that "the one idea of Western people is to have a town as good as New York." In their striving to be like New York, Cheyenne residents wanted the comfort and convenience of a big city, and so pursued the achievements of running water and electric lights. In the 1870s the Union Pacific Railroad provided the business district with piped water, while the residences relied on wells. For irrigation, the city filled ditches twice a week throughout the residential district. Finally, in 1883, the city's new water system was completed, providing piped water as well as a sewage system throughout the city. In 1882 city leaders began exploring electric lighting. They initially chose the Brush Electric Light Company, and by January 1883 thirty businesses were equipped with electric lighting. Technical problems as well as managerial incompetence contributed to Brush's failure to supply electricity for houses, however, and it was not until the fall of 1885 that Cheyenne's houses were adequately served by a modern electrical system.[45]

Nonetheless, the city of Cheyenne outstripped the army in its provision of conveniences, providing piped water to its citizens two years before the fort provided it to its personnel, a sewer system seven years before the fort, and electricity nineteen years before the fort. The army tried to provide comfort and convenience to its officers and men, often employing a moral argument when it could not otherwise justify the expense. Thus, even the provision of oil lamps earned a moral justification: it would keep the men out of bars. Clear medical reasons involving ventilation and sanitation provided impetus to build spacious barracks, water systems, and bathhouses, although flush toilets did not fall in this realm. In providing conveniences familiar to civilians, the army generally followed, rather than led, but it was every bit as concerned with trying to represent a civilized society. Like a village, army posts were not bastions of modernity and up-to-date conveniences, but they were comfortable and familiar places.

PART III | INSTITUTION
Fort D. A. Russell, 1890–1912

6

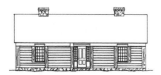

A MONUMENT TO THE PORK BARREL

Purpose, Plan, and Architecture

The 1890s were a turning point for forts in the West. The frontier had disappeared, as the secretary of war noted, "in the march of population and civilization westward." The Battle of Wounded Knee in 1890 represented the last armed resistance of the Plains Indians, ending four decades of warfare in the region. The concentration of forts that Sherman had initiated in the early 1880s had been achieved by 1890, reducing the number of posts by more than one-third, from 187 to 117.[1] Forts Laramie and Bridger both fell victim to these closures. At both forts, the army transferred troops to other sites, salvaged building parts for use at other posts, auctioned buildings and other property to the public, and turned the land over to the Department of Interior to be made available to homesteaders. A chapter in the history of the U.S. Army in the West closed.

But by the end of the decade, the U.S. military had a new purpose: international imperialism. The army more than doubled in size, requiring new facilities and new administrative structure. A modern, centralized military for the twentieth century replaced the sleepy western army of the 1880s. Fort D. A. Russell rode the crest of this wave, despite a disadvantageous location. With the help of a powerful politician, Fort D. A. Russell grew into a tangible representation of the New Army, writ in red brick.

After 1890 Fort D. A. Russell had no clear reason to exist. It was not near any ports or major transportation centers, nor was it near an Indian reservation, the last likely place for conflict in the West. Its survival and even growth over the next twenty-two years was due to the power of one politician,

Francis Emroy Warren. In a period when the army attempted to concentrate its forces at fewer forts and rationalize its programs, Warren managed to funnel $5 million to his pet project, Fort D. A. Russell. His actions defied logic and efficiency, angered army reformers, and earned him the reputation of a pork-barrel politician par excellence.

The monument that stood as a result of Warren's efforts was impressive indeed. By 1912 more than one hundred fifty brick buildings on tree-shaded streets housed some three thousand officers and soldiers. The size of the investment made it difficult for the army to abandon the fort, even without Warren's continued vigilance. Although products of old-style political maneuvering, the buildings represented new-style army efficiency. Built to standard plans issued in Washington, D.C., with all the work contracted to civilian firms and watched over by a construction supervisor, the new fort buildings were sound in construction, standard in plan, and modern in method.

Reorganization and Senator Warren

The old-style politician responsible for this rebuilt fort was a Republican who represented Wyoming in the U.S. Senate for thirty-seven years. Francis E. Warren was a special-interest politician in the Progressive Era, an easy target for reform-minded muckrakers. In a highly critical article published in *Collier's Weekly* in 1912, when Warren was up for reelection, C. P. Connolly leveled many charges, including undue influence over judges, coal-land fraud on behalf of the Union Pacific, bribery of newspaper editors and communities, provision of federal jobs to his relatives, fencing of public lands for personal use, holding of private contracts with the government, pasturing of his sheep at government expense, and tax dodging. Investigators found that Warren fenced public lands in 1886 and 1906; a House committee reiterated the charge in 1912.[2] Warren was, however, an extremely effective advocate for Wyoming, especially where Fort D. A. Russell was concerned. During his long career in the Senate, he chaired both the Military Affairs Committee and the Appropriations Committee, which enabled him to funnel federal funds directly to Wyoming.

Warren was born in Hinsdale, Massachusetts, in 1844. After service in the Union Army he moved to Cheyenne in 1868. Amasa R. Converse, a Hinsdale native, employed him in his mercantile business; Warren bought him out in 1878. During the 1880s Warren expanded his business interests, riding Cheyenne's wave of prosperity. Like many of the other wealthy men in Cheyenne, Warren had extensive livestock holdings; he owned seventy thousand head of sheep and probably an equal number of cattle. He invested in Cheyenne real estate, including several prominent commercial buildings, one of which he leased to the government for a post office. He also invested in local gas, telegraph, irrigation, electricity, railroad, and street railway companies.[3]

Warren's first political office was Cheyenne's city council, to which he was elected in 1883. The next year, he was elected mayor and was also elected to serve in the territorial legislature. In 1889 President Harding appointed him territorial governor; when Wyoming became a state the next year, he was elected state governor. Months later he ran for the U.S. Senate and was elected in 1890. Because he drew the short straw, he was up for reelection in 1892, and he lost. In 1894 the state legislature sent him back to the Senate, where he remained until his death in 1929. From 1905 until 1911 Warren chaired the Senate Military Affairs Committee, a position that enabled him to ensure Fort D. A. Russell's survival.[4] Warren's connection was clear: a year after he assumed the chairmanship, the secretary of war announced that Fort D. A. Russell would nearly triple in size to accommodate a brigade. The stream of construction funds for the fort came to a halt a year after Warren relinquished the chairmanship in favor of that of the Appropriations Committee, where he sent federal funds to Wyoming through other means.

There is no doubt that Warren engaged in pork-barrel maneuvering. Connolly's investigative article charged that Fort D. A. Russell "has cost the Government $5,000,000." Connolly also called the fort "a monument to the 'pork barrel,'" noting that "it is one of the largest military posts in the United States. Its population is one-fifth the size of Cheyenne."[5] But where Connolly saw this expenditure as a huge waste, the people of Cheyenne viewed it differently. Providing jobs for its citizens, markets for its products, and consumers for its entertainments, the fort was an undeniable boon to Cheyenne. Forts had always served as financial windfalls to western communities. Warren was particularly blatant about exploiting that advantage, and his constituents appreciated his efforts.

Despite Warren's effectiveness as a supporter of Fort D. A. Russell, his anti-Progressive actions were in direct opposition to trends in the army. During the same years when Warren piped money to the fort, the army underwent a comprehensive reorganization along professional lines. Sentiment within the army for reorganization had been building for some time, but without popular support vast reforms went unfunded. Nonetheless, before General Sherman resigned his position in 1883, he managed to begin the concentration of his troops at fewer but better forts.[6]

In 1898 the United States annexed Hawaii, invaded Cuba and the Philippines, and waged war on Spain, unveiling the new role of the United States as a world power. Imperialism proved popular with the American public, and Congress appropriated $50 million for national defense. The inefficiencies inherent in the army's rapid expansion and administration of territories became apparent during the war, with much of the blame falling on the secretary of war, Russell Alger. Reasoning that colonial administration was the most pressing issue, President McKinley replaced Alger with a corporation lawyer, Elihu Root. The Spanish-American War had also revealed weaknesses in the army's administration, including the mobilization of reserves and the structure of the command, as well as the lack of a strategic planning capability. Root soon realized that fundamental changes within the army were necessary, and he skillfully ushered them through Congress and past objections from some officers. In his five-year term in office, Root laid the groundwork for the New Army.[7]

Root rationalized the bureaucracy by eliminating the independent position of commanding general of the army, replacing it with a chief of staff who reported to the secretary of war. Most staff officers reported to the chief of staff. Where formerly the quartermaster general had reported to the secretary of war, not the commanding general, now the quartermaster general reported to the chief of staff. This reform proved difficult to implement over the objections of the current commanding general, Nelson Miles, so Root scheduled this reform to go into effect one week after Miles's retirement in 1903. The general staff was to plan for war, not administer or operate. Administration was left to the Adjutant General's Department, which was newly consolidated with the Record and Pension Office to form the Office of the Military Secretary. Recognizing that a standing army would never be adequate for complex foreign engagements, Root also worked to develop an expandable corps through the reserves and through the National Guard, which was brought under the army's supervision.

Root was concerned about the performance of the army in the field. In 1901 he received authorization to staff an army that could range between sixty thousand and one hundred thousand men. When the regiment was the organizational tool, companies of a regiment were dispersed among various posts, and the company was the basic unit. Now Root argued in favor of large posts, hoping to bring brigades (three regiments) together, combining the branches of infantry, cavalry, and artillery, as in a battle situation. Advantages included officers receiving experience in handling large bodies of troops, men being under the direct supervision of officers of high rank, potential comparison and emulation between officers and their units, and the education available at service schools that could be maintained at large posts. While calling life in small posts "narrowing and dwarfing," Root also recognized some benefits: the scattering of the army helped to popularize it, and the nation could achieve an equitable distribution of troops among states. Nonetheless, economy, efficiency, and military policy argued for concentration.[8]

In 1901 Root undertook a survey of posts. Senator Warren recommended that a battery of field artillery be permanently stationed at Fort D. A. Russell in addition to the battalion of infantry already there. Warren pointed out that "Fort Russell is a well-built post, healthy, convenient, with good water supply, sewerage, etc." He mentioned the target range, space for drills, pasturage, and stables. Only a "little inexpensive extension in men's quarters" would be required, he argued, as all other necessary buildings were there. Although Warren's arguments may not have been persuasive, his role as chair of the Military Appropriations Committee was, and the fort received continued funding. In 1906 Root's successor as secretary of war, William Howard Taft, recommended that the fort be enlarged to brigade size, or a capacity of more than three thousand men. The fort, which then accommodated one regiment of infantry, a squadron (four troops) of cavalry, and two batteries of field artillery, would receive additional cavalry and artillery troops to increase the post's strength to a regiment each of infantry, cavalry, and artillery.[9]

Six years later, Chief of Staff Leonard Wood, a proponent of Root's reforms and the New Army, urged further concentration of forces at fewer posts. A 1912 study undertaken by the War Department listed Fort D. A. Russell as one of the posts to be abandoned. A hint of this appeared a year earlier, when Wood assembled a full division of sixteen thousand men for maneuvers on the Mexican border. The troops from Fort D. A. Russell arrived six days late because of the "inability of the railroads concerned to furnish sufficient equipment . . . due to the isolated location of Fort D. A. Russell." In the 1912 report the fort's location was further criticized: "This post is not located with a view to maximum economy or strategic effectiveness. Its position in a sparsely settled region involves an increased cost for transportation of manufactured supplies, and its distance from recruiting centers makes the recruitment of its garrison more costly." Despite its excellent facilities and the $4,925,486.15 spent on the post to date, the report listed Fort D. A. Russell as one "which would have to be abandoned in order to put an end to the extravagance and inefficiency resulting from improper distribution of the mobile army."[10]

Wood faced opposition from Fort D. A. Russell's most ardent supporter, Senator Warren. The army appropriations bill for 1912, developed by Warren, authorized a commission of congressmen to report on the location and distribution of posts and also provided that no officer could serve as chief of staff unless he had served at least ten years as an officer of the line—legislation clearly designed to remove Wood, who did not meet this criterion. The president vetoed the legislation, but Wood's proposal to close forts was unsuccessful, and Fort D. A. Russell continued to exist. The post received no more large appropriations or ambitious construction programs until World War II, but it escaped elimination.

Purpose

As the Indian wars ended, Fort D. A. Russell's role shifted until there was no logical reason for the fort to survive. The various uses to which troops were put—or places to which troops were sent—illustrate this shift. By the 1890s much of the army in the West was stationed at posts near Indian reservations, such as Fort Washakie, Wyoming, near the Wind River Agency, or Fort Robinson, Nebraska, near the Pine Ridge Agency in South Dakota. Fort D. A. Russell was not close to any reservation, but it did have the advantage of being located along the railroad. In the 1890s the fort sent its troops by train to areas of conflict.

The last military action between the U.S. Army and American Indians is generally recognized as the Battle or Massacre of Wounded Knee, South Dakota, in 1890. Sioux Indians practicing the Ghost Dance religion sufficiently alarmed their reservation agents and neighboring settlers that the army was called in. Nine companies of the 17th Infantry from Fort D. A. Russell traveled to South Dakota by taking the train to Fort Robinson and marching north. A tense standoff, in which members of the 7th Cavalry attempted to disarm a group of 350 Sioux making their way to the Pine Ridge Agency from the north, resulted in an explosion of violence and the deaths of 146 Indians, including 44 women and 18 children.[11]

When Bannock Indians were reported active near Jackson Hole in 1895, Fort D. A. Russell troops again mobilized. Traveling by railroad to Market Lake, Idaho, and marching east, the troops found no confrontation or violence, but only a situation that had been enflamed by settlers.[12] As was often the case in the West, in the 1890s Anglos were far more likely than Indians to be the source of troubles requiring troops. The actions of Fort D. A. Russell troops in suppressing riots by Anglos against Chinese coal miners in Rock Springs in 1885 had foreshadowed this new role.

The Johnson County War was a conflict between small ranchers and big cattlemen. The latter, organized into the Cheyenne-based Wyoming Stock Growers Association, often suspected the small ranchers of stealing their cattle. In 1892 members of the Wyoming Stock Growers Association hired about fifty men to punish suspected rustlers in Johnson County in a vigilante action. The mercenary force, known as the Invaders, took the train from Cheyenne to Casper and headed north by horse. They killed two suspected rustlers at a ranch about fifty miles south of Buffalo, enflaming the local populace, who sympathized with the small ranchers. The Invaders holed up at a ranch about seven miles south of Buffalo, where the sheriff and a large deputized force laid siege to them. Notified of this predicament, Acting Governor Amos Barber sided with the Stock Growers Association and wired the president of the United States to request assistance. Senator Warren allegedly rousted Benjamin Harrison out of bed to authorize troops. Soldiers from Fort McKinney, west of Buffalo, rescued the Invaders, arrested them, and delivered them to Fort D. A. Russell, where they were held pending trial. The Johnson County prosecutor eventually dropped the charges, and the Invaders were set free.[13]

A nationwide depression in 1894 contributed to domestic unrest. That year, when Commanding General John M. Schofield noted that there was "domestic violence in half the States," Wyoming was not immune. Members of Coxey's Army, a collection of unemployed people responding to a populist call for a march on Washington, commandeered trains in Wyoming. When the U.S. marshal recaptured a train in Green River and arrested fifteen leaders, troops from Fort D. A. Russell arrived to prevent "interference with the property of the Union Pacific Railway by Coxeyites, Commonwealers, tramps, et al." Soldiers escorted 147 prisoners to Boise, Idaho. When railroad workers in Wyoming went on strike in sympathy with the strikers at Pullman, near Chicago, troops were stationed at several points to maintain the peace.[14] While such duty did little to endear the army to the general public, big businessmen and politicians found the army increasingly useful.

Troops from Fort D. A. Russell were also involved in the foreign engagements that caused the army's popularity back home to surge. Beginning in 1898 troops from the fort rotated in and out of Cuba and the Philippine Islands, and after 1911 to Mexico.[15] Wyoming was not an advantageous location for any of these stations, and was even less so for the looming intervention in Europe. Considerable new construction during this period, however, meant that Fort D. A. Russell would be difficult to abandon.

Layout

When in 1911 a journalist criticized the "elaborate suburban villages" that forts had become, he might have been thinking of Fort D. A. Russell.[16] Between 1890 and 1912 the army added more than one hundred fifty buildings to the post, transforming it from a village to a small city. The quarters grew in size, the construction was of high quality, and the surroundings were increasingly pleasant. The post grew linearly, extending

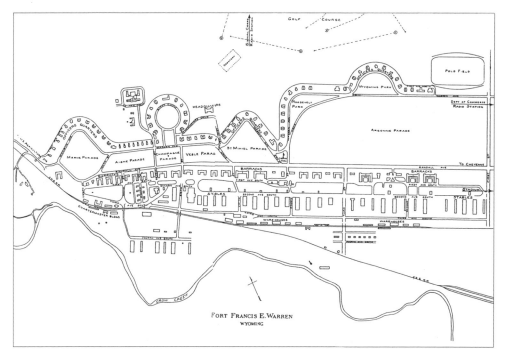

Fig. 6-1. Fort D. A. Russell, site plan, Cheyenne Chamber of Commerce, 1930. The original diamond-shaped parade ground is on the far left. The meandering line of officers' quarters increased their frontage on the parade ground. Courtesy F. E. Warren Air Force Base.

east from the original diamond in a trend begun with 1880s construction (fig. 6-1). The elongated parade ground, stretching more than a mile to the east, nearly to the reservation boundary, separated officers from enlisted men. On the south side of the parade ground, aligned barracks faced onto the parade. Behind them stood a row of stables, then a row of warehouses. South of them were quarters for noncommissioned officers. Officers' quarters lined the north side of this enlarged parade ground. Behind them were a hospital, a headquarters building, baseball field, polo fields, and target ranges. In 1906 the post housed a regiment of infantry, four troops of cavalry, and two batteries of field artillery, comprising about eight hundred men.

Once Secretary of War William H. Taft announced, at the end of 1906, that Fort D. A. Russell would become a brigade post, a new alignment was overlaid on the old. While construction still moved to the east, preserving the north-south zones, an east-west grouping fell slowly into place: regiments of infantry on the west, cavalry in the center, and field artillery on the east, with officers' quarters facing their respective barracks. This recalled the original arrangement of the fort, with cavalry officers and enlisted men on the east side of the diamond and infantry officers and enlisted men on the west side. A loop of officers' quarters called Staff Circle indicated that these housed officers not attached to regiments. Here, then, was Root's vision of a brigade organization articulated on the landscape: the three branches in the same fort in separate zones, maintaining a rigid hierarchy between officers and enlisted men.

Standardization

Despite the change in plans necessitated by the designation of a brigade post, the buildings themselves maintained a degree of uniformity throughout the 1899–1912 building program. New buildings took their cues from existing ones. All red brick, they borrowed from comfortable classical and colonial revival styles, then popular across the nation, with exterior and interior symmetry, white trim, columns, and a spacious feel. By 1906 the quartermaster department used the colonial revival style for its standard designs, except for posts in the Southwest, where it employed a Spanish colonial style in a nod to regionalism.[17] The seeming architectural variety at Fort D. A. Russell derived from the mix of ranks, not from any planned variation. Each rank of officers received similar buildings, but because the ranks were intermingled, a pleasing variety struck the observer. Despite the domestic quality of the varied façades and the colonial revival style, the size of the new buildings, especially the barracks, put them far outside the domestic realm, as the new construction took on an increasingly institutional face.

Part of this institutional quality was due to the use of standard plans, which were produced by the Quartermaster General's Office. The first appearance of this standardization at Fort D. A. Russell occurred with the design of the 1894 administration building (fig. 6-2). Located at the south point of the original diamond, the two-story brick building followed plans issued a year earlier by the Quartermaster General's Office. Labeled "Administration Building at [blank]," the drawings did not have a plan number, but were clearly designed for use at multiple posts. The center hall of the first floor was flanked by offices for the commanding officer, adjutant, and sergeant major on one side, with offices for clerks on the other. The side-gable roof with parapeted end chimneys, splayed stone lintels with keystones, and round-arched windows in the second floor of the gable ends were features not found on any other buildings at Fort D. A. Russell.[18]

Little construction occurred in the next five years, but in 1899 several major changes took place. First, the wood-frame officers' quarters, dating from the establishment of the post in 1867, were removed, solidifying Fort D. A. Russell's new image as a permanent, brick-built post. Second, foreseeing

Fig. 6-2. Fort D. A. Russell, administration building, ca. 1894. The workmen are probably from Black & Clark, the Cheyenne firm that received the contract to construct this building in 1893–1894. Courtesy Wyoming State Archives, Department of State Parks and Cultural Resources.

considerable new construction over the next several years, the Quartermaster's Department assigned a superintendent to the post, thereby finally professionalizing construction supervision. F. J. Grodavent, who had worked at Fort Stevens, Oregon, and Fort Logan, Colorado, arrived in Cheyenne to begin work at a salary of $1,500.[19] The appearance of a construction superintendent meant that the post quartermaster no longer had to oversee complex construction jobs for which he had no training.

Third, in another sign of professionalization, plans for nearly every building constructed came from the Quartermaster General's Office and were stamped with a standard plan number. This move to standardize architecture, in which plans issued from Washington were used at different posts, went into effect with little discussion or notice. The standard plans simplified contracting. Plans and specifications developed by the Quartermaster General's Office were accurate; if not, they were revised in subsequent issues. Similarly, designs were refined through experience. The Quartermaster General's Office issued subsequent editions of the designs, which received letter designations appended to plan numbers, e.g., 142-F.[20] The change revolutionized the duties of the post quartermaster. Once a new building was authorized and funds designated for its construction, the chief quartermaster sent the standard plans and specifications to the post quartermaster, who then advertised for bids. The chief quartermaster selected the lowest bidder, and the superintendent of construction oversaw the work. Once construction was complete, the superintendent of construction accepted the building from the contractor and turned it over to the post quartermaster.

Standardization also revolutionized the appearance of army posts nationwide. A cursory examination of some of the posts near Fort D. A. Russell reveals great similarities in layout and general appearance. Fort Mackenzie near Sheridan, Wyoming, Fort George Wright near Spokane, Fort Robinson near Crawford, Nebraska, Fort William Henry Harrison near Helena, Montana, Fort Lincoln near Bismarck, North Dakota, and Fort Crook and Fort Omaha, both near Omaha, were all built or rebuilt between 1890 and 1912. Maintaining the general hierarchy of officers' quarters and barracks, with noncommissioned officers' quarters off the parade ground and hospital separate from the rest, the buildings were all red brick and in interchangeable designs. Officers' quarters, barracks, stables, guard houses, and hospitals all had an eerie familiarity. While all were laid out around a parade ground, none of the parade grounds was rectilinear; all of them were pulled into acute angles or horseshoe curves or other irregularities. The repetition of the architecture from post to post put an institutional stamp on these varied places, identifying them as *army* buildings.[21]

Besides more efficient designs and streamlined contracting, the army also believed standard plans would help control costs. The Quartermaster General's Office intended to produce designs that were cheaper to build than what might be produced at the post, and contracting construction to the lowest bidder would ensure savings. But these good intentions clashed with demands for more luxurious and spacious officers' quarters and with rising costs of materials. Caught between the ideal and real was Quartermaster General Charles F. Humphrey, by coincidence the same man who, as the depot quartermaster, oversaw much of Fort D. A. Russell's rebuilding in the late 1880s.

Humphrey enlisted as a private in the artillery during the Civil War, then began his ascent through the ranks, transferring to the Quartermaster Department in 1879 with the rank of captain. Between 1885 and 1891 he commanded the Cheyenne Quartermaster Depot and supervised construction at Fort D. A. Russell. By 1896, when he was posted to Washington, he had served the previous twenty-two years continuously stationed west of the Missouri River. He then served as chief quartermaster in Cuba, China, and the Philippines until Elihu Root appointed him quartermaster general with the rank of brigadier general in 1903. Root wanted someone who had not been holding a "soft" staff job in Washington, and Humphrey was his man. Humphrey also had the backing of Senator Warren. Entirely self-taught, Humphrey explained in his 1890 efficiency report that he had "an excellent library" and that "special study for the last ten years has been on the subject of construction of buildings for military purposes; heating and water supply for same; construction of water and sewerage systems; construction of roads, etc. etc." Humphrey would retire in 1907 after forty-four years of army service.[22]

Humphrey's first problem as quartermaster general was the flood of work that fell upon his office in the first decade of the twentieth century. Humphrey did not have the manpower to handle an appropriation for construction and repair of about $9 million per year, more than ten times the usual appropriation in the late nineteenth century. He recognized that he could have assigned some of the work to chief quartermasters or post quartermasters, but he "kept within this office the work which could be most advantageously, economically, and properly handled here," in other words, the development of standard plans and specifications. If he had decentralized this work, the effects would have been "objectionable," "resulting in want of uniformity of plans, productive of endless correspondence and discussion, and removing the work from the immediate supervision of the Secretary of War." In 1906 he received an additional 33 clerks to supplement the 191 who were already on his payroll.[23]

Humphrey's next problem was the escalating cost of materials. In 1905 he observed that construction costs had risen 36 percent in the past three years, and the estimated increase for 1906 was 12 percent. The cost of lumber was the greatest concern, as it doubled between 1904 and 1907. Accordingly, the army began using reinforced concrete construction in an effort to minimize the use of wood. By 1910 the quartermaster general was able to report a modest reduction in construction costs.[24]

Savings also derived from changes in designs. In January 1905 Humphrey hired Francis B. Wheaton, the first architect on the army's staff. Wheaton had worked in the office of the supervising architect of the treasury, the office that had most federal construction under its control, for the previous ten years. He had also worked for the prominent architectural firms of McKim, Mead & White and Van Brunt & Howe. In the army, he was charged with revising the standard plans and specifications to improve "the general appearance of buildings" and to eliminate "unduly elaborate details." He did this in part by the application of a more rigorous colonial revival style, one that was plainer and had a simpler plan. Among the specific cost-saving changes mentioned were the following:

Omission of plaster on inside walls of barracks wherever practicable and substitution of paint placed directly on brick or concrete, with incidental reduction in cost of wood trimming, thus saving not only in first cost but also in cost of upkeep from year to year; toilet rooms in barracks have been reduced in size and material savings made in eliminating some of the expensive slate partitions heretofore used; in plans of officers' quarters the interior has been rearranged by eliminating waste spaces, thus permitting decrease in size without material reduction in accommodations; exterior trimming of buildings has been modified with a view to reduction in cost; specifications have been revised so as to require only material of standard dimensions in general use, thus obviating the necessity for articles of special forms, dimensions, and grades.

Standard materials naturally resulted from the standard specifications issued by the Quartermaster's Department, but in the case of plumbing fixtures the Quartermaster's Department went outside its domain and devised plumbing specifications in accordance with the Treasury and Navy Departments. The quartermaster general recognized that this might have nationwide ramifications: "All of the manufacturers expressed the opinion that the commercial world will largely adopt the fixtures called for in these specifications."[25]

The goal of regularizing construction costs at different posts through the use of standard plans was not always achieved. Fort Mackenzie, a new post in northern Wyoming, received its first funding (thanks to Senator Warren) in 1900. In 1905 construction of the same design of double captains' quarters cost $20,611 at Fort D. A. Russell and $26,473 at Fort Mackenzie. Similarly, two-company barracks at Fort D. A. Russell cost $60,444.90 in 1905 but $76,044.00 at Fort Mackenzie a year earlier. Even at the same post, costs varied. Artillery barracks built in 1904 cost $48,010.67, while six months later another set cost $46,267.96. In 1908 Congress established limits on construction costs of quarters at permanent seacoast defenses: $15,000 for brigadier general quarters, $12,000 for field officers, and $9,000 for company officers, including plumbing and heating. No quarters had been built for these amounts or less, even at seacoast fortifications; fortunately, Congress did not extend these limitations to inland forts. Nonetheless, the quartermaster general began preparing plans that would fit within these new guidelines, and applying them to the "mobile army" as well. Secretary of War Luke Wright recognized that "a limitation upon the cost of quarters is in accordance with correct business principles," aligning the army with modern ideas of business management.[26]

Standard plans also proved advantageous to contractors, who procured materials and constructed buildings. Bidding for a job could be based not just on general experience, but on the experience of constructing a particular building type, either at the same fort or a different one. Several contractors competed for the construction of Fort D. A. Russell in the early twentieth century, including Russell Sanders Construction Company, but much of the work went to one man—Moses Keefe. Based in Cheyenne, Keefe obtained contracts at several forts in the West, using his experience at one to prepare bids at another.

Keefe, born in Ireland in 1853, came to the United States at the age of seventeen. He first went to Decatur, Illinois, then worked construction in Chicago after the fire. He came to Wyoming, attracted by tales of the battle at Little Big Horn, and worked at Fort McKinney and Fort Laramie as a civilian builder in 1877. He established a contracting business in Cheyenne in 1879, which he operated until 1914. Locally, he built both residential and commercial buildings, including the State Capitol, which he cited in his advertisements. Active politically, he served in the territorial legislature in 1886, as a county commissioner, as a city councilman, and as mayor in 1902 and 1903. He obtained construction contracts at Fort Crook, Nebraska, in the 1890s and then participated in the rebuilding of Fort D. A. Russell in the early 1900s. He spent a year and a half in Cuba beginning in January 1899, also under military auspices. He built extensively at Fort George Wright, near Spokane, using his experience at Fort D. A. Russell to build to some of the same standardized plans. His biography credited him with building "half of Fort Russell" and claimed that "he is a man wholly self-made, possessed of fine executive ability, and will always 'deliver the goods.'"[27] It is also clear that he was a man who did very well for himself through the "privatization" of military construction.

As Built

The stage was set, then, for the institutionalization of Fort D. A. Russell. Standard designs from Washington, a professional construction supervisor and civilian contractors, and an emphasis on efficiency framed the early twentieth-century architecture of the fort. Officers' prerogatives were expressed in the scale of their quarters rather than in the individuality of their post.

The officers' quarters revealed several hierarchies at work. They occupied one side of the parade ground; accommodations for the men stood on the other. The enlisted men lived in barracks; officers, in houses. Among officers, quarters reflected rank. Lower ranking officers received double houses, while higher ranking officers had single, freestanding quarters. Rank also earned more rooms. The space allocation of the 1850s was still

Fig. 6-3. Fort D. A. Russell, bachelor officers' quarters (Building No. 21), ca. 1905. The large building accommodated unmarried officers and was built to a standardized plan issued by the Quartermaster General's Office, QMGO Plan 129. Courtesy F. E. Warren Air Force Base.

in effect in 1900, but was observed less and less. However, the number of rooms was still a mark of rank, including rooms intended for servants. All officers' quarters included servants' quarters, and the plans showed a variety of ways of accommodating them in a separate circulation pattern. A second set of stairs in the back of the house gave access to servants' spaces on all floors: the laundry in the basement, the kitchen on the main floor, sometimes a bathroom on the second floor, and bedrooms in the attic. Occasionally a door would define the back hallway on the first and second floors, keeping the servants out of sight. The hierarchy here was not only in the number of servants that a higher ranking officer would be able to house, but also in the extent to which he and his wife could keep the servants separate from their family life.

The first of the standard buildings, a bachelor officers' quarters (BOQ) completed in 1900, signaled another change (fig. 6-3).[28] With its very name, the BOQ indicated that unmarried officers were the exception, not the rule. Previously, officers' quarters had been assigned by rank, regardless of family. Now, single officers received smaller quarters, housed together in

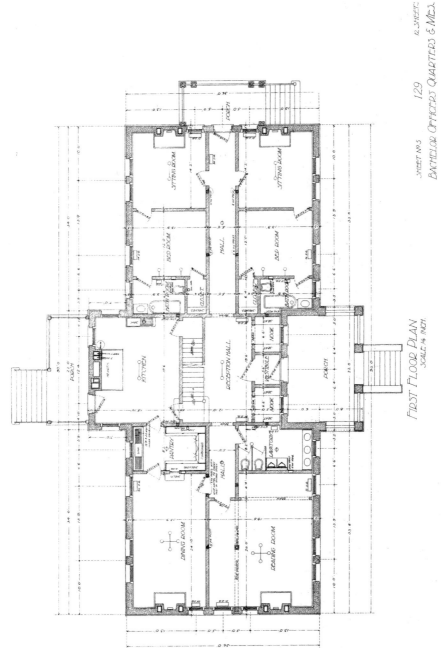

Fig. 6-4. QMGO Plan 129, first-floor plan, 1899. Standardized Plan 129 indicates that its use was not limited to Fort D. A. Russell. Courtesy F. E. Warren Air Force Base.

one building. The new BOQ was located in a prominent position, at the turn between the old diamond and the newer officers' quarters built in the 1880s. The plans, which bore a number from the Quartermaster General's Office, provided for six suites, each consisting of a sitting room, bedroom, and bathroom (fig. 6-4). The common areas included a kitchen, dining room, reading room, billiard room, and reception hall with two nooks. The 90-foot-long building looked markedly different from neighboring houses, but it was officers' quarters nonetheless.

In the fall of 1900 the chief quartermaster let contracts for more buildings: two double officers' quarters, intended for lieutenants; one double officers' quarters, intended for captains; and a guard house. On November 14, contracts went to M. P. Keefe for construction, Codgess Construction Co. of Chicago for electrical wiring, and Johnson & Davis of Denver, Dwyer Plumbing & Heating of St. Paul, and P. S. Cook of Cheyenne for plumbing and heating. The lieutenants' quarters followed plans issued by the Quartermaster General's Office and also used for quarters at Fort Mackenzie, Fort Lincoln, North Dakota, Fort Myer, Virginia, and Fort Ethan Allen, Vermont (fig. 6-5).[29] The design called for asymmetrical, mirror-image plans with the entrances in the end bays, recessed 7 feet from the front façade. The first floor included a reception hall, parlor, dining room, and kitchen, while the second floor had three bedrooms and a bathroom. A common porch stretched across the front. A central cross gable, two bays wide, was flanked by portions of the main roof that extended to the porch roof in a rather odd arrangement.

The double captains' quarters built at the same time had a similar arrangement; the entrance bays on the ends were set back from the front of the building (fig. 6-6). The reception hall was wider, the dining room larger, and a back set of stairs was added. Separate fireplaces in the dining room and parlor replaced corner fireplaces sharing the same chimney in the lieutenants' quarters (fig. 6-7). The captains' quarters had four bedrooms and a bathroom on the second floor, as well as a functional third floor, which contained two bedrooms and a bathroom for servants. Each front entrance had its own porch; paired, round-arched windows in the cross-gable and gabled dormers lit the attic. The building was built to the same plans as officers' quarters at a dozen other posts.[30]

The difference between lieutenants' quarters and captains' was subtle, as both ranks inhabited double houses. The additional space provided the captain meant another bedroom, more room for servants, a grander reception hall, and larger public rooms. The common wall, though, served as a reminder that the officer occupied only half of a house, compromising privacy. While the sounder construction of these brick buildings prevented the kind of eavesdropping possible in earlier, more porous double houses,

Fig. 6-5. Fort D. A. Russell, lieutenants' double quarters (Buildings Nos. 18 and 20), ca. 1905. Built in 1900–1901, these double quarters had a common front porch and slightly recessed entrances. Courtesy F. E. Warren Air Force Base.

there were nonetheless areas of conflict. One way to minimize confrontations between neighbors was to manipulate the front entrances and porches. The double officers' quarters built in 1867 had entrances flush with the front of the building and a common porch across the front (fig. 6-8). This was changed in the 1900 lieutenants' quarters to entrances recessed from the front, and a common porch. In the 1900 captains' quarters, the entrances were recessed, and there were separate porches. The next evolution, lieutenants' quarters built in 1909, had entrances on the sides and separate porches.[31] With the main entrances on the sides, the quarters provided for a back-to-back arrangement in which the adjoining unit was much less noticeable.

Officers' quarters became increasingly elegant and spacious. The addition of a third floor providing two bedrooms and a bathroom for servants, as well as a back stairway for servants and more closets on the second floor, signaled this in the lieutenants' quarters. For captains' quarters, a new set of standard plans issued in 1900 included an additional public

Fig. 6-6. Fort D. A. Russell, captains' double quarters (Building No. 12), ca. 1905. Also built in 1900–1901, double quarters for captains had separate front porches and additional space inside. Courtesy F. E. Warren Air Force Base.

room on the first floor, to serve as a small office or library. This provision of more rooms foreshadowed a change in room allowances, finally granted in 1908, in which each rank gained at least one more room as quarters.[32]

Beginning in 1905, and undoubtedly coinciding with Wheaton's employment, captains' quarters reflected the use of a simpler colonial revival style. The 1905 version was much blockier than the earlier, without the cross-gable, and its façade was in one plane (fig. 6-9). Further, flat-arched windows in the later building replaced segmental- and round-arched windows of the earlier one. The plan was rationalized, with a squarish reception hall replaced by a linear one (fig. 6-10). With changes such as these, the architecture reflected the current mode while being cost-conscious.

These officers' quarters typified new construction at Fort D. A. Russell over the first decade of the twentieth century (fig. 6-11). The fort erected more than sixty officers' quarters, of which at least forty were double houses for lieutenants and captains; the rank of major and above earned a single dwelling. Majors commanded battalions of infantry and squadrons

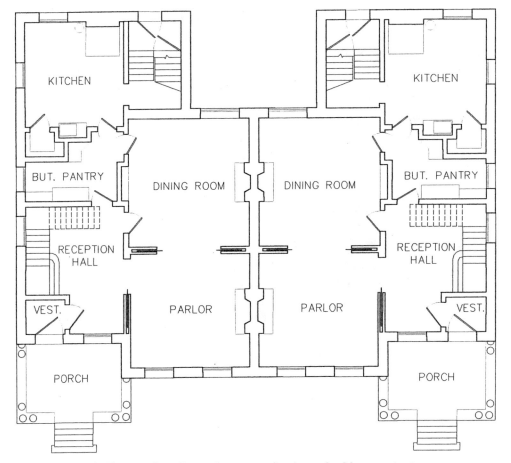

Fig. 6-7. QMGO Plan 90, first-floor plan, 1894. In these double captains' quarters, more than a quarter of the space on the first floor is devoted to servants. From Grashof, "Standardized Plans, 1866–1940"; redrawn by author.

of cavalry, while lieutenant colonels were usually second-in-charge of regiments; the consolidation of larger units at the brigade post meant that more senior officers had to be accommodated. These single field officers' quarters were spacious, without necessarily adding any more rooms.

In 1902 Fort D. A. Russell was a regimental post headed by a colonel. Moses Keefe received the contract to construct a commanding officer's quarters at a cost of $11,500, completing it in early 1904.[33] This new quarters was located near the north point of the original diamond, next to the site of the first commanding officer's quarters (figs. 6-12 and 6-13). The Quartermaster General's Office had issued the plans in August 1897, and the quarters was clearly more a nineteenth-century building than a twentieth. It had an asymmetrical side-hall plan, a three-story bay window on one side,

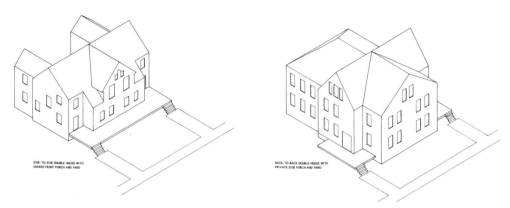

Fig. 6-8. Fort D. A. Russell, axonometric view of two officers' quarters (Building No. 18–20 and 127–128). On the left is the double lieutenants' quarters built in 1900–1901 with a common porch (also depicted in fig. 6-5). On the right is the double lieutenants' quarters built in 1909, with entrances in the sides and separate porches for more privacy. Drawing by Barrett Powley, 1996, Western Regional Architecture Program, Graduate School of Architecture, University of Utah.

and round-arched windows with brick hoodmolds in the front gable. The paired white columns on the porch, the denticulated cornice, and the red brick with white trim related this building to the rest of its contemporary construction. The entrance opened into a generous stairhall; occupying the other front corner of the house was a parlor, joined by sliding doors to the dining room. On the other side of the hall was the "office and library" and in the rear, the kitchen. On the second floor, four rooms were labeled "chamber"; there was also a sizable sewing room, a full bathroom, and a room with just a toilet. In the attic were three bedrooms and, surprisingly, two bathrooms.

With the fort's designation as a brigade post in 1906, additional colonels as well as a brigadier general needed housing. In 1909 two quarters were built to plans from the Quartermaster General's Office that had been designated for a commanding officer (fig. 6-14).[34] At Fort D. A. Russell, these were assigned to the colonels of the cavalry and artillery regiments. The former was located at the top of the half-diamond off the extended parade ground, opposite the cavalry barracks, and the latter was placed at the top of a half-circle, opposite the artillery barracks. These buildings had a much more straightforward colonial revival appearance, with a five-bay façade, pilastered dormers, and a rigid symmetry. The central entrance led into an 8-foot-wide hall, with the library to one side and parlor and dining room on the other. The house strictly separated family and servants; servants used the rear hall and stairs, while the family had its front hall and stairs. The second floor had four bedrooms and two bathrooms. The "servants'

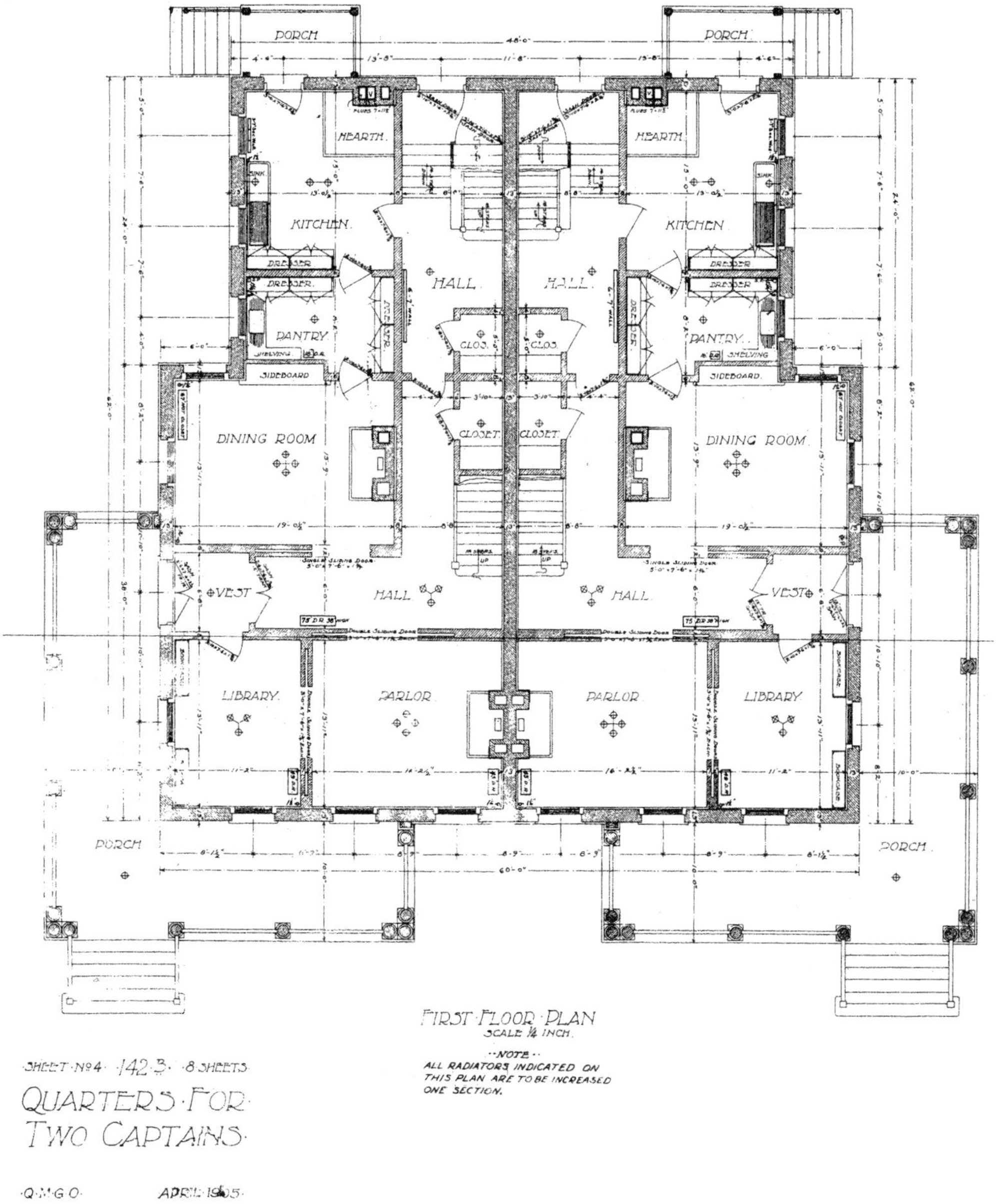

Fig. 6-9. QMGO plan 142-B, first-floor plan, 1905. The standardized plan of the captains' quarters has a linear hall, rather than the reception room in QMGO Plan 90, as well as a library. Courtesy F. E. Warren Air Force Base.

bath" was located off the rear stairs. On the third floor, three bedrooms in the front of the house shared a bathroom, while the two other bedrooms, presumably for servants, opened onto the rear stairs.

By July 1909 Fort D. A. Russell had finally attained brigade status, and Brig. Gen. Fred A. Smith arrived to take charge of the post. The Quarter-

Fig. 6-10. Fort D. A. Russell, officers' quarters, photo by J. E. Stimson, ca. 1910. The first and third buildings from the left are double captains' quarters and the fourth and fifth are double lieutenants' quarters (also seen in fig. 6-5), all built in 1905. Second from the left is single quarters for a field officer, while on the right, in the distance, is bachelor officers' quarters constructed in 1908. Courtesy Wyoming State Archives, Department of State Parks and Cultural Resources.

master General's Office issued plans for quarters for the brigadier general but gave them a "special" number, indicating that they were not widely used. Nonetheless, Fort Monroe, Virginia, also received quarters to this plan. Fort D. A. Russell's example was completed in 1910.[35] Located at the corner of the enlarged parade ground, the house had an expansive view and a commanding presence, today augmented by an *allée* of trees that focuses attention on the house; the quarters was designed to see and be seen (fig. 6-15). The commanding officer's quarters was clearly intended for public functions. The center doorway opened into a wide reception hall, more than 24 by 15 feet, with a fireplace and two nooks. The parlor and dining room, each 16 by 25 feet, flanked the hall. A relatively cozy library, 12 by 15 feet, was tucked into the rear wing, which also contained service spaces of kitchen, laundry, and rear stairway. Seven bedrooms and

Fig. 6-11. Fort D. A. Russell, officers' quarters, ca. 1910. First, fourth, and fifth from left are captains' quarters in the simplified Colonial Revival mode (also seen in fig. 6-9). Between them are double lieutenants' quarters. The two-story portico belongs to the commanding officer's quarters, which is flanked by single quarters for field officers. Courtesy Wyoming State Archives, Department of State Parks and Cultural Resources.

three bathrooms occupied the second floor, while the third floor had two bedrooms and a bath.

But it was on the outside that the grandeur of this house was really apparent. A two-story portico with paired Ionic columns dominated the front façade. A terrace stretched across the front of the building. The double doors of the entrance were crowned with an elliptical fanlight; above it, on the second floor, floor-to-ceiling windows opened onto a balcony. The house had brick quoins, some stone trim, and a modillioned cornice. The gleaming white columns against the brick wall announced the building as one of importance. It succeeded in being the grandest at the post.

It was also the grandest in Cheyenne. Comparison with another very important residence in town, the governor's house, was instructive. The governor's mansion was built in 1904 to designs by Charles Murdock of Omaha (fig. 6-16). It, too, had a two-story portico, although with paired Corinthian columns. It, too, was designed for entertaining, with a large parlor (20 by 27 feet) and dining room (28 by 16 feet). But the wider front

Fig. 6-12. Fort D. A. Russell, commanding officer's quarters (Building. No. 8), ca. 1905. In 1904, when the fort was a regimental post headed by a colonel, this commanding officer's quarters was built to QMGO Plan 95-A. Courtesy F. E. Warren Air Force Base.

of the commanding officer's quarters (63 feet compared to 55) and the correspondingly wider portico give the military building the greater presence. In a comparison of two residences built for public display and entertainment, the army's grand and imposing building is equal, if not superior, to the governor's mansion.

Just as officers' quarters reached new extremes of spaciousness and luxury, some grumblings were heard among the officer corps. One officer suggested that due to the difficulty of obtaining servants, the army should provide light-housekeeping flats for younger or childless married officers. An officer's wife proposed bungalows as a cheaper and easier alternative. Col. William S. Patten, assistant quartermaster general, disagreed, arguing that "the average Army officer wants about all the house and as much *porch* as he can get with his rank."[36] Realistically, officers would select the largest house to which they were entitled, regardless of the housekeeping burden.

Quarters for noncommissioned officers (NCOs) fit awkwardly into the established dichotomy of officers vs. enlisted men. With barracks on one

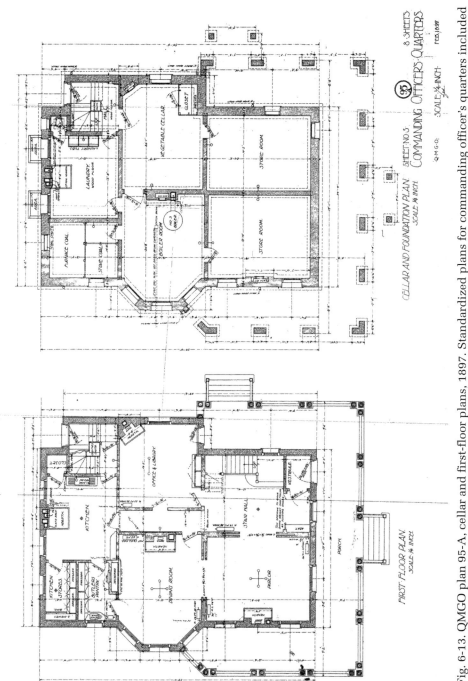

Fig. 6-13. QMGO plan 95-A, cellar and first-floor plans, 1897. Standardized plans for commanding officer's quarters included a bay window in the dining room. Courtesy F. E. Warren Air Force Base.

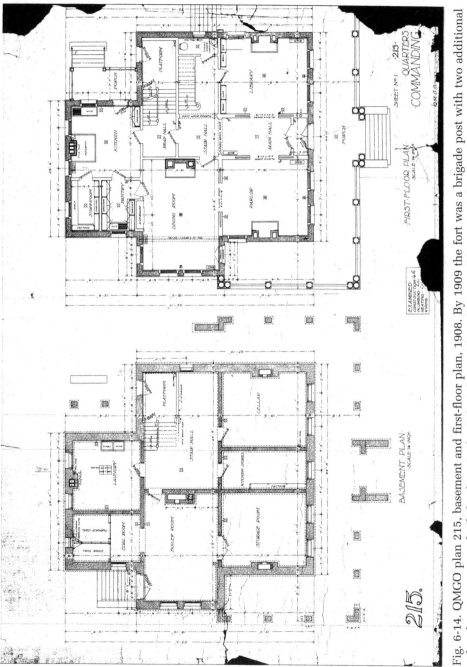

Fig. 6-14. QMGO plan 215, basement and first-floor plan, 1908. By 1909 the fort was a brigade post with two additional colonels; quarters were built for them in the Colonial Revival mode dictated by QMGO Plan 215. Courtesy F. E. Warren Air Force Base.

Fig. 6-15. Fort D. A. Russell, commanding officer's quarters (Building No. 92), photo by J. E. Stimson, ca. 1910. The post's brigade status brought it a brigadier general, who required even grander quarters, built in 1910. Courtesy the Stimson Collection, Wyoming State Archives and Historical Department.

side of the elongated parade ground facing officers' quarters on the other, NCO quarters remained an oddity. In 1885, with the increase in the number of post NCOs, a line of NCO quarters was placed south of the barracks and perpendicular to the expanding direction of the fort, which was east-west. These six quarters were one-story cottages, each measuring about 22 by 27 feet, with one bedroom. In 1908 two sets of double NCO quarters were built facing them, on the same street. In the next two years six more double houses were built for NCOs, but far to the east, and on an east-west orientation.[37] These newer houses were also far to the south; they stood south of the barracks aligned on the parade ground, south of a line of stables, and south of a line of warehouses. Like the barracks and officers' quarters to the north, the NCO quarters observed the zones, with infantry NCOs housed in the western grouping, and cavalry and artillery NCOs on the east.

The eight new NCO quarters were built to two versions of the same plan, which provided two units of two stories with mirror-image floor plans, each measuring 18½ by 27 feet (fig. 6-17). Entrance in the outside

A MONUMENT TO THE PORK BARREL ■ 231

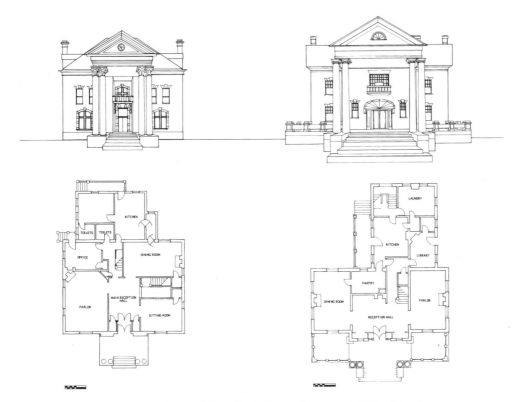

Fig. 6-16. Governor's mansion and Fort D. A. Russell commanding officer's quarters, perspective drawing. While both the governor's mansion and the commanding officer's quarters were porticoed houses with large rooms, suitable for entertaining, the general's house had grander presence. Drawing by Heather Randall, 1996, Western Regional Architecture Program, Graduate School of Architecture, University of Utah.

bay led to a side hall containing a straight-run stairway. A "sitting room" and kitchen occupied the first floor, while two bedrooms and a bathroom were on the second. The houses were snug, but larger than the 1885 NCO quarters. The disadvantage of the new quarters was that they were double houses, with a common porch across the front.

Although very few buildings were constructed at Fort D. A. Russell in the two decades between the wars, most of the buildings that were built were NCO quarters. The army added eleven double houses in the early 1930s, on both sides of the road on which the 1910 quarters had been built. They bore great similarity to the 1910 houses, being brick double houses, most with side-gable roofs. But the new houses, which shared the same plan, were larger: each unit was about 21 by 30 feet, allowing a third bedroom to be placed on the second floor, and a corner fireplace in the living room. Each unit also had a one-story sun porch off to the side. The new houses were varied in exterior appearance, with hip and gable roofs and different entrances. The separate entrances were in the center bays of

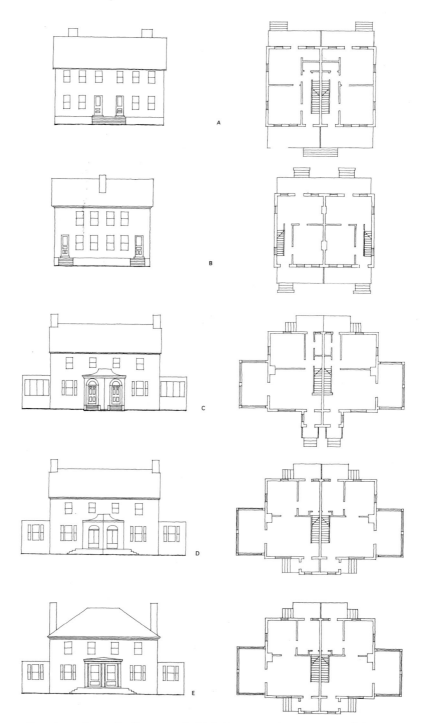

Fig. 6-17. Fort D. A. Russell, diagram of NCO Quarters. Six quarters of the type B in this diagram were constructed in 1908–1910 on the southern edge of the inhabited area of the post. Type A, a variant of that standardized design, was doubled into a four-unit building for civilian employees, constructed in 1911. Types C, D, and E are variations of another set; eleven were built in the 1930s. Drawing by Liza Hart, 1996, Western Regional Architecture Program, Graduate School of Architecture, University of Utah.

Fig. 6-18. Fort D. A. Russell, barracks, photo by J. E. Stimson, ca. 1910. From right to left, infantry barracks for two companies built in 1905, the two artillery barracks built in 1904, and two sets of cavalry barracks for two troops, built in 1906. Courtesy Wyoming State Archives, Department of State Parks and Cultural Resources.

the buildings: some were entered from the front, some from the side; some had round-arched doorways, some flat-arched; some were brick, some wood. The variations, though modest, nevertheless gave some dynamism to the street and some sense of individual identity to the houses.

Barracks lined the south side of the parade ground, along the main entrance road (fig. 6-18). Unlike the north side of the parade ground, the south maintained a relatively straight alignment. With the barracks visible in one glance, their collective size projected a powerful image. Between 1903 and 1912, when construction halted, thirteen large barracks were built to three different designs. Their varied appearance had less to do with planned variety than with the vagaries of construction funding.

Two barracks for artillery were completed first, in 1904 (fig. 6-19).[38] The two-story barracks was in a U-shaped plan with two-level porches across the 140-foot front. A cross-gable roof crowned the projecting two-story entrance bay. On one side of the main hall were a day room, store room, offices, and, in the rear wing, a dormitory that slept two dozen men. On

Fig. 6-19. Fort D. A. Russell, two-company barracks for artillery (Building No. 231–232), ca. 1908. Built in 1908 to QMGO Plan 181, this barracks had a porch that did not stretch all the way across the front, unlike its neighbors. Courtesy F. E. Warren Air Force Base.

the other side of the hall were the mess room and, in the rear wing, the kitchen. On the second floor, two L-shaped open dormitories and some smaller rooms flanked the center hall. Six barracks shared another standardized design for two companies in which projecting pavilions at each end denoted the separate entrances for the companies.[39] Five more two-company barracks, built to a third standardized plan, stretched 200 feet across their fronts (fig. 6-20).[40] On these, the projecting end pavilions were more noticeable, as the two-level porches did not extend across them (fig. 6-21). All of the sleeping space was located on the second floor in two dormitories, and the kitchen, mess room, and day room, along with various offices, were on the first floor (fig. 6-22). The basement held the bathroom, barber and tailor shops, and the boiler room.

Once brigade status was planned in 1906, the barracks assignments began to accord with the zoning.[41] Infantry companies occupied all of the older barracks from the 1880s to the west, as well as the two barracks built for them in 1905. The Signal Corps occupied the two barracks built for artillery in 1905. Cavalry troops occupied the next six large barracks to the east, and artillery batteries the easternmost three. Within these zones, and also facing onto the parade ground, were additional buildings: administration buildings for infantry, cavalry, and artillery; guard houses for infantry, cavalry, and artillery; band barracks for cavalry and artillery; and post exchanges for cavalry and artillery.

The post exchange and gymnasium for the infantry was located farther south, off the parade ground (fig. 6-23). Built in 1905, the building offered two innovations for enlisted men. The first was the post exchange, a club for soldiers that provided refreshments and amusements, turning the profits back to the men. The second, the gymnasium, was the outgrowth of a fitness

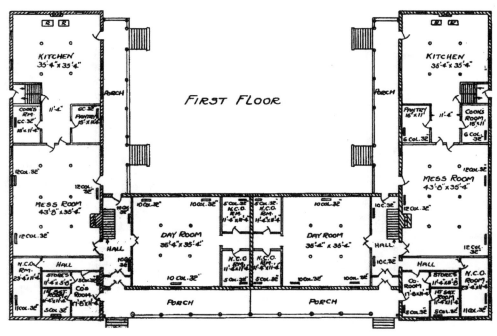

Fig. 6-20. Fort Hancock, N.J., two-company barracks, first-floor plan, 1930s. The barracks built to QMGO Plan 181 were two bays shorter but otherwise the same as Fort D. A. Russell's. Courtesy National Archives.

craze of the 1880s. In 1896 the commanding general ordered that the men exercise daily for thirty minutes at least nine months of the year.[42]

The post exchange and gymnasium at Fort D. A. Russell was a two-story, T-shaped building with brick pilasters and a modillioned cornice. In the rear wing, the gymnasium offered vaulting bars, climbing ropes, parallel bars, punching bags, rowing machines, and other equipment, all of which could be swept away to make room for a basketball court (fig. 6-24). Above was a suspended running track. The basement contained a bowling alley, shooting gallery, lunch room and kitchen. In the main block, the post exchange and lockers occupied the first floor, while the second floor had a lecture room, reading room, and classroom. Such complete facilities for the enlisted men reflected a subtle change in attitude about their value and what they deserved.

The zoning continued behind the barracks, in an impressive range of twenty-two brick stables, in which cavalry stables were clustered behind the cavalry barracks, and artillery stables, interspersed with gun sheds, stood behind the artillery barracks. Quartermaster stables were located on a different alignment, farther south, behind the infantry barracks. At its largest, the standard stable plan measured 67 by 220 feet and was designed

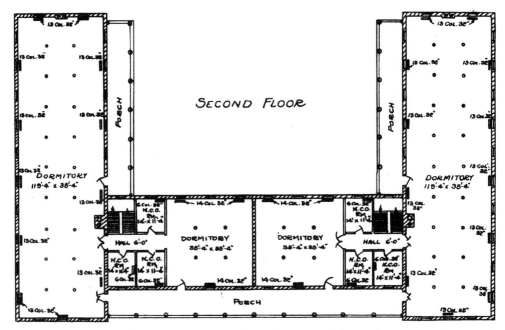

Fig. 6-21. Fort Hancock, N.J., two-company barracks, second-floor plan, 1930s. The second floor was devoted to dormitory space. Bathrooms were in the basement. Courtesy National Archives.

to hold 122 horses, appropriate for a battery of artillery, while a troop of cavalry required accommodation for only 62.[43] All variations contained 5-foot-wide stalls on the perimeter, lit by small windows. Between two longitudinal corridors were a saddle or harness room, box stalls, and additional regular stalls. A clerestory monitor roof lit and ventilated the interior. The substantial construction devoted to horses occurred just as their utility was ending; the mechanized army would appear in the next decade.

Landscaping

Efforts to achieve a grassy parade and well-shaded streets continued. In 1904 the commanding officer requested grass seed for 20 acres. The next month, the chief quartermaster refused to sign off on 750 feet of garden hose and twelve lawn sprinklers, believing they were inappropriate to the military mission. A year later Senator Warren urged that trees be planted along the road from the fort to the edge of the reservation, which was the boundary of Cheyenne, and the post complied. The source of the water for irrigating grass and trees continued to be a problem, as the city of Cheyenne persisted in disputing Fort D. A. Russell's right to take water from the pipe that crossed the military reservation. In 1902, the army refused to proceed

Fig. 6-22. Fort D. A. Russell, barracks, ca. 1910. A rare interior view, this party scene in an artillery barracks shows pressed-metal ceilings and electric light fixtures. Courtesy Wyoming State Archives, Department of State Parks and Cultural Resources.

with new construction unless the controversy was resolved. A new contract in 1903 finally settled the issue for irrigation water. In 1908 the post signed another contract with the city that ensured its supply of household water. When the point of diversion from Crow Creek was changed to Crystal Lake, the army paid $400,000 to the city as its share in the construction of reservoirs, pipelines, and the Round Top Filter Plant.[44]

Another twentieth-century addition to the parade ground was a streetcar line. In 1908 the Cheyenne Electric Railway, of which Senator Warren was a part owner, ran from downtown Cheyenne, out to the post, and then 3 miles onto the post to the edge of the original diamond-shaped parade ground. Providing easy access to downtown, the streetcar should have been a boon to the soldiers, but instead its fares proved to be a bone of contention. A soldiers' boycott in 1909 succeeded in temporarily reducing the fare from eight cents to five. In January 1911 the soldiers again began campaigning for a reduction in the fare. They held a boycott, first using any other available means to get downtown. Then they decided to take their business

Fig. 6-23. Fort D. A. Russell, post exchange and gymnasium (Building No. 284), photo by J. E. Stimson, ca. 1910. Providing amenities to enlisted men, this building, constructed to QMGO Plan 158, had a prominent modillioned cornice, two-story pilasters, and a pedimented doorway. Courtesy the Stimson Collection, Wyoming State Archives and Historical Department.

elsewhere. In September the Fort Russell Progressive Club, an organization of soldiers, arranged an expedition to Denver via train, and succeeded in transporting 720 men there for a twenty-four-hour leave. They pledged to do so every payday until the Cheyenne Electric Railway Company lowered its fares. Finally, a year after the boycott began, the railway offered tickets for purchase at five cents at the post exchange, while the cash fare remained ten cents.[45]

Furnishings and Conveniences

By the time of Fort D. A. Russell's early twentieth-century building spurt, most of the conveniences were in place. All new quarters and barracks had running water and flush toilets. All new buildings had steam heat. All were wired for electricity, which was available after the fort hooked up to

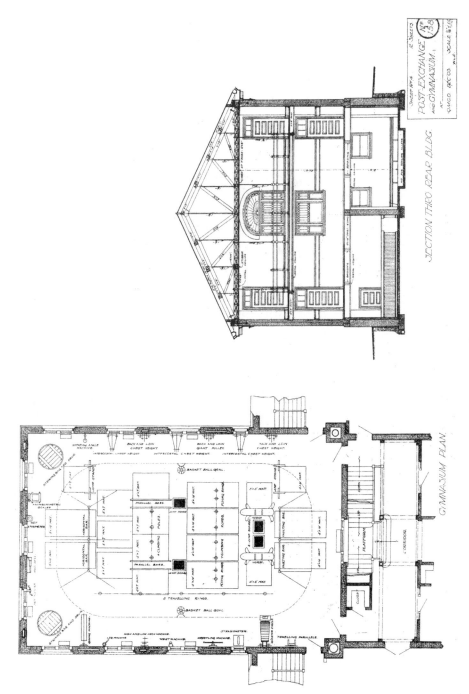

Fig. 6-24. QMGO Plan 158, gymnasium plan and section through rear building, 1903. The gymnasium floor indicates the myriad activities that could occur there. Courtesy F. E. Warren Air Force Base.

Cheyenne's utility, in which Senator Warren had a controlling interest, in 1904. Fort D. A. Russell was not far behind the norm in the provision of electricity to military posts. In 1908 the quartermaster general reported that 79 posts were lit by electricity, 7 by gas, 2 by acetylene gas, and 64 by kerosene lamps. Six years earlier Brig. Gen. Frederick Funston, commanding the Department of the Colorado (in which Fort D. A. Russell was situated at that time), argued in favor of electricity, using the comparative method: since "all public buildings owned by various departments of the Government, transports and naval vessels, and almost all towns and villages in the vicinity of outlying posts are lighted by electricity, it seems that the Army is very antiquated in its lighting of military posts."[46] Electricity, at least for Fort D. A. Russell, followed soon after.

One army-wide provision of comfort in this period concerned furniture. Responding to years of complaints, the army in 1907 agreed to provide heavy furniture for officers' quarters. Although the Grand Rapids Furniture Company developed samples, J. M. Shellenburg of Philadelphia obtained the contract to provide more than $200,000 worth of furniture. Fort D. A. Russell received 54 dining room tables, 306 side chairs, 102 armchairs, 65 desks, and 18 sideboards, all in mahogany. By 1911 the army also provided bookcases, chests of drawers, a divan, a parlor table, hall tree (coat rack), range, and refrigerator, along with instructions for the care of mahogany furniture. Noncommissioned officers, much to their chagrin, received only a range and a refrigerator.[47]

Ideas about sanitation had advanced since John Shaw Billings struggled with the issue in the 1870s. In 1909 Major P. M. Ashburn of the Medical Corps issued *The Elements of Military Hygiene*, which was immediately adopted as a textbook in the post schools. Ashburn, like Billings, emphasized ventilation, but recognized that colds and such were caused by microorganisms, not miasms. In the barracks, he advocated large squad rooms and a minimum of 600 cubic feet of space and 60 square feet of floor space per man. Experience in the tropics had taught the value of screens, especially in the barracks and mess hall. By 1905 the quartermaster general provided door and window screens for barracks, quarters, and hospitals, specifically to prevent the spread of disease, as well as to make the occupants of the building more comfortable. Ashburn claimed that showers and flush toilets were installed in almost all posts, and he instructed that men should bathe at least twice a week in summer and once a week in winter. Throughout, Ashburn emphasized cleanliness, particularly in the barracks, mess hall, and kitchen, in which "the cook . . . should live, think, and dream cleanliness." Such ideas about hygiene coincided with less expensive construction as well as a sleeker, more modern look in architecture. Thus, the inspector general announced in 1904 that wood wainscoting was omitted from new barracks "to prevent the harboring of vermin." Similarly, enameled steel wall lockers replaced wooden wardrobe lockers in 1905; they were also cheaper, due to the high cost of lumber.[48]

It is not certain if the improved surroundings actually made the men any happier, but desertions declined. In the 1870s and 1880s the annual desertion rate was about 15 percent; in the first decade of the twentieth century, it was about 5 percent. Changes in enlistment regulations in the 1890s had altered the makeup of the troops, shifting the balance toward American-born. Immigrants had made up about half of the enlisted men in the first decade after the Civil War; that proportion had dropped to about one-third in the 1880s. But in 1894 Congress required that recruits had to be able to read, write, and speak English and that they had to be citizens or declare their intent to become citizens. Because of the depressed economic circumstances of that decade, the army easily filled its ranks with literate, American-born recruits, and by 1910 immigrants made up only 10 percent of the force.[49]

Perhaps because soldiers were more likely to be native-born, their circumstances received greater attention, especially from muckraking journalists. One such article, entitled "The Shame of Our Army: Why Fifty Thousand Enlisted American Soldiers Have Deserted," by Bailey Millard, deplored the "high" desertion rate and attempted to find the cause. Accommodations were not a factor; Millard described barracks at Fort Snelling, where nearly one-third of one company deserted, as "new, sanitary, and comfortable." Nor was it the rigors of weather, as most soldiers deserted in the summertime. Instead, the author pointed to the tasks to which soldiers were assigned: "to dig ditches, wash pots and pans, wait on table, clean out stables, sweep off walks, or cut brush in the hot sun." The more onerous jobs of constructing buildings and infrastructure had been contracted out, but still fatigue duty was seen as unreasonable. Millard also pointed to the endless drills in bad weather and "the drudgery and monotony of garrison life," a complaint that had been heard earlier as well.[50] The architectural improvements army-wide at the turn of the century were given little credit for improving morale.

One group with an extremely low desertion rate—less than half of one percent—were the African American troops. African Americans had begun serving in the army during the Civil War and had filled four regiments after the war, stationed at various posts in the West. Their desertion rates were consistently lower and their reenlistment rates consistently higher than the rest of the army, indicating either a general satisfaction with army life or, more likely, a lack of opportunity elsewhere. The blacks were kept in segregated companies, commanded by white officers; they inhabited the same barracks that whites did, at different times. Four companies of the 24th Infantry, the first African American troops assigned to Fort D. A.

Russell, arrived in September 1898 from action in Cuba. In 1902 Troop E of the 10th Cavalry, another black unit, arrived at Fort D. A. Russell from assignment in the Philippines.[51] The African Americans caused little stir in the community.

The officer corps also saw some changes in this period. What had once been ostensibly a life of leisure had now become one of physical exercise, administrative duties, and studying, due to the army's new emphasis on officer training. Officers' wives lived, by one account, a modest life, "what with the increased cost of living and the unchanged figures on the monthly pay check." Army life called for a positive attitude, a lack of interest in frills, and housewifely skills that could stretch a budget. It provided "unique little communities where still exists a manner of life so simple, so wholesome, and withal so gracious."[52] As in the past, officers' wives were put on a pedestal and admired for their ability to endure hardship without complaint. The hardship, though, was no longer danger and deprivation but, rather, fewer material goods than others of their class.

The Modern Fort

Standardization of the architecture, involving a rationalization of the building process resulting in greater efficiency and potentially greater economy, was attributable to national ideological currents of the day. Fundamentally, standardization required the recognition of professionalism. Commanding officers and post quartermasters, once reluctant to concede authority to the Quartermaster's Department in Washington, finally acknowledged that the expertise of those professionals was necessary. Effective hierarchies, such as that spanning staff and line, which entailed the recognition of professionalization, contributed to a centralized administrative power. In that period, the bureaucracy had great appeal nationwide as a remedy for the social and political upheavals caused by industrialization. Modernization, the general concession to organized, cosmopolitan ideas, thus propelled the trend toward professionalization and standardization.[53] In the army, the bureaucracy triumphed over a disparate network of disorderly posts, organizing these far-flung construction enterprises through standard designs.

Standardized buildings illustrated the ideology of modernization. Identical buildings dotted the military landscape, not only at Fort D. A. Russell, but at other posts throughout America, lending a cohesion and uniformity to army construction as a whole. Army-wide standardization created a monolithic institution, so that the similarities of the posts translated into a strength of purpose and military might. The buildings were of sound brick construction, the officers' quarters were spacious, upper-class dwellings, the barracks were enormous and imposing, the open space was generous. A single pedestrian was dwarfed by the expanse of open space and by the size and number of the buildings. As Fort D. A. Russell lost its individuality in this period, it gained a surer foothold in the larger army. Through its monolithic presence, the fort expressed its role in the development of the United States into an imperialistic world power.

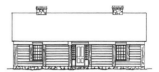

EPILOGUE

The history of these three forts did not end with the privatization of Forts Laramie and Bridger in 1890 or with the cessation of a vast building program at Fort D. A. Russell in 1912. Today, Forts Laramie and Bridger are important tourist destinations that present the western past to generations of curious travelers. Fort D. A. Russell is an active U.S. Air Force base that preserves its historic architectural core, which lends an air of dignity and authority to a high-tech military mission. Although none of these forts experienced extensive new construction in their historic areas after the time considered here, their history in the twentieth century, related briefly, indicates an ongoing connection to their nineteenth-century past.

Once Senator Warren gave up the chairmanship of the Senate Military Affairs Committee in favor of the chairmanship of the Senate Appropriations Committee, his efforts on behalf of Fort D. A. Russell had more to do with fighting off periodic threats of closure than obtaining funds for new construction. Warren's successful infusion of federal funds into the fort before 1912 ensured its survival in a time when the army was seeking to close superfluous posts. But for the next thirty years, there was very little new construction at the fort. In 1930 the post was renamed after Warren, who died in November 1929, still a senator from Wyoming.[1]

In the 1930s new construction was not extensive. A barracks and four officers' quarters built alongside their respective types were very similar in appearance and faced onto the parade ground. Two buildings, a theater and a gymnasium, were built in the open space of the parade ground. Eleven noncommissioned officers' quarters were added to the southern grouping of NCO quarters, and a medical barracks was constructed near the hospital. On the eve of World War II, Fort Francis E. Warren looked very much as it had on the eve of World War I.

During World War II, a great deal of new construction occurred at Fort Francis E. Warren, but most of it was off the original parade ground. In 1941, south of the original fort and south of Crow Creek, a Quartermaster Replacement Training Center for basic training of Quartermaster Corps draftees was constructed of temporary buildings, 282 built in six months. By the time the United States entered the war, 387 buildings there could accommodate 20,000 men.[2] Most of the Replacement Training Center was demolished in the 1960s.

After the war, the fort was turned over to the newly established U.S. Air Force, which in 1949 renamed it Francis E. Warren Air Force Base. In the early 1950s a housing development on a curvilinear street plan was constructed on the eastern edge of the fort (fig. E-1). Five hundred Wherry houses, named after the sponsor of the legislation for their construction, were built. Government-subsidized, the houses were built and managed by a consortium of Cheyenne builders. The ranch-type homes were very unlike those of the old Fort D. A. Russell, but also very separate from it. In 1958 the base was assigned to the Strategic Air Command, which established an Atlas Intercontinental Ballistic Missile headquarters there. New family housing construction continued; in 1962 about ninety houses in the Capehart Housing Project were built northwest of the original fort. Soon after Atlas construction ended in 1962, F. E. Warren was selected as the support base for 200 Minuteman missile sites, 50 of which were replaced in 1986 with Peacekeeper missiles.[3] About half of the Wherry housing units were demolished in 1988, and new housing was constructed south of Crow Creek.

The heart of the base retains its historic character and in 1975 was designated a National Historic Landmark Historic District. The Air Force takes pains to maintain the historical integrity of its historic buildings, and in recent years nearly all new construction has taken place outside of the historic core (fig. E-2). This historic center of F. E. Warren Air Force Base remains a monument to Fort D. A. Russell's greatest period of development, the 1899–1912 building boom. The immense growth during that period is important not just for its quantity, tripling the size of the base, but also for the triumph of standardization that these buildings represent.

When the army abandoned Forts Bridger and Laramie as a cost-saving measure, jurisdiction shifted to the Department of the Interior, which was

Fig. E-1. F. E. Warren Air Force Base, Wherry Housing, ca. 1951. New housing constructed after World War II differed in form and layout from the pre-1912 officers' quarters in the foreground. Courtesy Wyoming State Archives, Department of State Parks and Cultural Resources.

Fig. E-2. F. E. Warren Air Force Base, officers' row, 1996. The curving road, spacious porches, and mature trees lend an air of civilian domesticity to the former Fort D. A. Russell. Photo by Thomas Carter, 1996, Western Regional Architecture Program, Graduate School of Architecture, University of Utah.

in charge of distributing them as public lands. In both cases, people already in residence there homesteaded the properties. The various owners farmed their land and used fort buildings for livestock or hay storage, or as their residences.[4]

The first interest in the forts as historical sites related to their role as landmarks on the Oregon Trail. Ezra Meeker, an Oregon Trail veteran, undertook a nationwide campaign for the marking of the trail, which resulted in the placement of 150 monuments by 1908. At Fort Laramie, one property owner, John Hunton, obtained the help of the Daughters of the American Revolution to build a concrete obelisk. Its granite plaque read:

FORT LARAMIE
A MILITARY POST ON THE OREGON TRAIL
JUNE 16, 1849–MARCH 2, 1890
THIS MONUMENT IS ERECTED BY
THE STATE OF WYOMING AND
A FEW INTERESTED RESIDENTS[5]

The monument's allusion to the Oregon Trail as defining Fort Laramie's significance was important. The reference connected the site to the Americanization of the West and implicitly related Fort Laramie to settlers moving westward in fulfillment of the nation's Manifest Destiny. The Oregon Trail also tied in nicely to the linear movement of tourists, who might be tracing by automobile their ancestors' journey. As a landmark on the Oregon Trail, Fort Laramie provided, then and now, a stopping point for travelers.

In 1925 Addison Smith, a U.S. representative from Idaho, introduced legislation to have the Oregon Trail designated a federal highway. Probably to support this effort, Wyoming's state legislature requested that the federal government "set aside Old Fort Laramie and Old Fort Bridger and Independence Rock as Historic Reserves." Although the federal legislation went nowhere, the state established the Historical Landmark Commission in 1927 as an agency that could condemn and receive properties, then turn them over to the federal government. Fort Bridger was the commission's first purchase, made in 1929 for $7,100.[6]

Fort Bridger's association with famed mountain man Jim Bridger was a compelling aspect of the fort's past. In a list of the Historical Landmark Commission of Wyoming's sites in 1930, Fort Bridger was described as "one of the most famed of trading posts on the western frontier." But the site that the Historical Landmark Commission acquired in 1929 contained no remnant of Bridger's fort, and only a portion of the stone wall that the Mormons built. Most of the surviving structures were army buildings, predominantly the stone ones built in the 1880s.[7]

In the thirty years the Historical Landmark Commission administered Fort Bridger, it concentrated on maintaining the site, placing signs in front of buildings, and establishing a museum for the display of artifacts in the post trader's complex. In 1959, with the abolition of the Historical Landmark Commission, the fort passed to the care of a state agency, the State Library, Archives, and Historical Board. Its initial project was to restore the one surviving log officers' quarters. In 1970 a house built from materials salvaged from the 1880s frame officers' quarters, but not resembling them in any way, was moved from its site about a mile north of the fort onto the southern side of the fort. A few years later, the 1884 commanding officer's quarters, which had been moved off the post in 1906 and dismantled in the 1940s, was reconstructed on its original site using the salvaged materials.

In 1984 the most significant addition to the fort occurred: the reconstruction of Bridger's trading post. With funding from the state legislature, contributions from the Friends of Old Fort Bridger, and supervision by the Archives and Museums Historical Department, the American Mountain Men Association built a reconstruction of the trading post at two-thirds scale, to the regret of museum professionals. The trading post, which is not on the original site, serves as the focal point of an annual Mountain Man Rendezvous, a four-day gathering for mountain men reenactors that draws more than twenty-five thousand participants, and illustrates, however imperfectly, an important aspect of Fort Bridger's past to many visitors.[8]

Fort Laramie, meanwhile, received the attention of the federal government, with the assistance of the state. When the associate director of the National Park Service, Arthur E. Demaray, said publicly in 1936 that the Park Service was "extremely interested" in preserving the fort, Wyoming governor Leslie A. Miller realized that the Park Service's participation was contingent on the state acquiring the property. The Historical Landmark Commission of Wyoming acquired the property, then deeded it to the U.S. government, and President Franklin D. Roosevelt proclaimed it Fort Laramie National Monument on July 16, 1938.[9]

Fort Laramie's historical significance has been well understood but rarely articulated. The proclamation designating the Fort Laramie National Monument noted only that "the lands and structures are of great historic interest," without defining that interest, and subsequent legislation has been no clearer. The site's association with the Oregon Trail spanned the trading-post phase as well as the army occupation, but, as at Fort Bridger, only the military era was represented by standing structures.[10] Since 1938 the National Park Service has meticulously restored some buildings and stabilized the ruins of others. Unlike the situation at Fort Bridger, it has not reconstructed any missing buildings.

As forts, the historic sites of Laramie and Bridger lack dramatic appeal. Never the site of battles, and having no association with a famous general like Custer, these forts are mostly known for their presence on the overland trail than their army occupancy. Yet the buildings that remain are almost entirely military structures. They represent less the military might of the army than its indirect but equally effective actions of subsidizing the developing western economy, and serving as a model and reminder of civilization left behind in the eastern part of the country.

Forts Laramie, Bridger, and D. A. Russell are all in the public domain, in some sense: Laramie as a national park, Bridger as a state park, and D. A. Russell as a U.S. Air Force base with a National Historic Landmark District. Administrators at all three sites wrestle, to differing degrees, with questions of significance as they decide on treatment of buildings. Officials at Laramie and Bridger examine their forts' pasts to determine the most appropriate course of interpretation to the public, which in turn guides their decisions about stabilization, restoration, and reconstruction. D. A. Russell, while less concerned overtly with presentation to the public, faces a constant stream of preservation decisions for buildings that are occupied and adapted to twenty-first-century uses. These decisions are governed by the fort's historic importance as set forth in the National Historic Landmark nomination form, as applied by various state and federal officials.

The buildings at Forts Laramie, Bridger, and D. A. Russell were powerful agents in the process of civilizing the West. Beyond their role as a base of military force, the forts served as landmarks, shelters, and places of opportunity for countless emigrants, settlers, visitors, and neighbors. In an alien environment of uncertain political domain, unknown personal safety, and tenuous economic existence, the forts functioned as a representative of the known. Here was an understandable social hierarchy, here was payment in U.S. dollars, here were canned goods and New York newspapers, and most of all, here were buildings that looked like those back East. With their domestic picket fences and bay windows, the orderly center-hall plans of officers' quarters, and their village green–like parade grounds, these sprawling communities instructed all who passed through what a respectable American settlement looked like. Village-like forts set an example, proving that it was possible to establish such a settlement, hundreds of miles from the nearest steamboat or railroad. The example was not followed literally, and neighboring communities often excelled forts in the provision of amenities, but for several decades forts provided both reassurance and hope that "civilization was still westward bound."

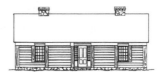

NOTES

INTRODUCTION

1. The best summaries of fort building are Robinson, *American Forts;* Frazer, *Forts of the West;* Prucha, *A Guide to the Military Posts of the United States, 1789–1895;* and Hart, *Old Forts of the Far West.* For a broader look at the western army, see Utley, *Frontiersmen in Blue* and *Frontier Regulars.* The social history of soldiers in the nineteenth-century West is explored in Coffman, *The Old Army;* and Rickey, Jr., *Forty Miles a Day on Beans and Hay.* While much information about the soldiers' day-to-day lives can be gleaned from their buildings, that is not the primary purpose here.

2. Strahorn, *Fifteen Thousand Miles by Stage,* 15.

3. See Robinson, *American Forts,* for a comprehensive history.

4. *Annual Report of the Secretary of War* (1882), 5; *Annual Report of the Secretary of War* (1883), 46; Smith, *The View from Officers' Row,* 133, 182; Bederman, *Manliness and Civilization,* 23; Utley, *Frontiersmen in Blue,* 42; Utley, *Frontier Regulars,* 410; Frazer, *Forts of the West,* xviii.

5. Hedren, *Fort Laramie in 1876,* 242.

6. Cronon, *Nature's Metropolis;* Limerick, *The Legacy of Conquest,* 28; Dobak, *Fort Riley and Its Neighbors;* Miller, *Soldiers and Settlers;* Smith, *The U.S. Army and the Texas Frontier Economy, 1845–1900,* 176; Ryan, *Fort Stanton and Its Community, 1855–1896,* 143; Priscilla Merriman Evans, "Pulling a Handcart to the Mormon Zion," in Marcus and Burner, eds., *America Firsthand,* 1:170; Fort Laramie Post Return, March 1867, Fort Laramie National Historic Site; Connolly, "Senator Warren of Wyoming," 10.

7. Limerick, *The Legacy of Conquest,* 27. Explorers of cultural conquest include Chris Wilson, in *The Myth of Santa Fe,* who studied the cultural contestation involved in creating and celebrating the history of a place; William H. Truettner, in the exhibit and catalog *The West as America,* who suggested that paintings of the West were designed to further the economic interests of the East; and Douglas C. Comer, in *Ritual Ground,* who analyzed the Americanizing effects of the rituals of the trading post. Scholars who have examined the power of architecture to express ideology include Henry Glassie, *Folk Housing in Middle*

Virginia; Bushman, *The Refinement of America;* Fred B. Kniffen, "Folk Housing: Key to Diffusion," in Upton and Vlach, eds., *Common Places;* and Dell Upton, "The Power of Things: Recent Studies in American Vernacular Architecture," in Schlereth, ed., *Material Culture,* 64.

8. Bushman, *The Refinement of America,* xii. Quantity of forts obtained from a rough count of those located west of the Mississippi, listed in Frazer, *Forts of the West;* and Prucha, *A Guide to the Military Posts.* Usually, there were between one hundred and two hundred western forts in operation at any one time, but the cycle of abandonment and new construction puts the total number of forts west of the Mississippi between 1844 and 1895 at more than three hundred. *Army and Navy Journal* 5 (11 January 1868): 330; Carrington, *Absaraka, Home of the Crows,* 141.

9. Prucha, *A Guide to the Military Posts,* pl. 19, and Frazer, *Forts of the West,* 179, both identify seventeen Wyoming forts, although their lists are slightly different. Histories of the three forts treated in this book are Hafen and Young, *Fort Laramie and the Pageant of the West, 1834–1890;* Lavender, *Fort Laramie and the Changing Frontier;* Gowans and Campbell, *Fort Bridger;* and Adams, *The Post near Cheyenne.* See these works for more information on the military, as opposed to the architectural, aspects of these forts.

CHAPTER 1. AN OASIS IN THE DESERT

1. Cited in Hafen and Young, *Fort Laramie and the Pageant of the West, 1834–1890,* 155.

2. Hafen and Young, *Fort Laramie,* 154, 148, 164. The register recorded 39,506 men, 2,421 women, 609 children, 9,927 wagons, and more than 75,000 head of livestock, according to Hafen and Young, who note that probably only four-fifths of the travelers were reported, and that there was probably a mistake in that number of children.

3. Holtz, "Old Fort Kearny, 1846–1848," 44–55.

4. White, *"It's Your Misfortune and None of My Own,"* 90. The threat of bodily harm was exaggerated, particularly on the Great Plains. Unruh calculated that of the 362 killings of Anglo-Europeans by Indians on the emigrant trails between 1840 and 1860, 90 percent occurred west of South Pass, making the first half of the journey westward by far the safer. The Indians were more intent on taking livestock than on killing emigrants. Nonetheless, the emigrants, backed by popular sentiment, demanded military protection for their routes. Unruh also contrasts the 362 killings of Anglo-Europeans by Indians with the 426 killings of Indians by Anglos. Unruh, *The Plains Across,* 144.

5. Utley, *Frontiersmen in Blue,* 113–117.

6. Simonin, *The Rocky Mountain West in 1867,* 86; Lavender, *Fort Laramie and the Changing Frontier,* 151–153.

7. Hafen and Young, *Fort Laramie,* 280, 197, 201, 254–255, 260, 261.

8. Utley, *Frontiersmen in Blue,* 281, 296, 197; Wooster, *The Military and United States Indian Policy,* 105; White, *"It's Your Misfortune,"* 96.

9. Sherman to Cooke, 11 September 1866, RG 393, U.S. Army Continental Command, Part I, Entry 3731, Box 1, National Archives Building, Washington, D.C. (hereafter NAB).

10. *Annual Report of the Secretary of War* (hereafter *ARSW*) (1868), 3; Clyde A. Milner II, "National Initiatives," in Milner, O'Connor, and Sandweiss, eds., *The Oxford History of the American West,* 181. In addition, the Sioux retained the right to hunt in the Powder River country.

11. *ARSW* (1857), 4–5, and (1853), 6; Utley, *Frontiersmen in Blue,* 55, citing *ARSW* (1854), 5.

12. *ARSW* (1869), 30.

13. *ARSW* (1872), 36.

14. *ARSW* (1869), 31.

15. Utley, *Frontiersmen in Blue*, 12, 14–17; Utley, *Frontier Regulars*, 11–14, 16; Wooster, *The Military and United States Indian Policy*, 30. Authorized company strength ranged between 37 and 100 privates.

16. *ARSW* (1858), 796.

17. *Cheyenne Daily Leader,* 22 July 1876; Hazen to Adjutant General, Department of the Platte, 29 August 1866, RG 393, Part I, Entry 3731, copy in vertical files, Sheridan County Fulmer Public Library, Sheridan, Wyoming; Murray, *Military Posts in the Powder River Country of Wyoming, 1865–1894,* 29.

18. Mackay to Jesup, Quartermaster General, 1 November 1849, RG 92, Office of the Quartermaster General, Entry 225, Box 531, NAB.

19. Mackay to Jesup, 1 November 1849; Holtz, "Old Fort Kearny," 52–53. Although Mackay's letter was dated November, he was at Fort Laramie at the end of July.

20. Hafen and Young, *Fort Laramie*, 179–182; Simonin, *The Rocky Mountain West in 1867,* 8.

21. Gowans and Campbell, *Fort Bridger,* 85.

22. *Cheyenne Daily Leader,* 26 August 1876.

23. Grimes cited in Oliva, "Fort Union and the Frontier Army in the Southwest," 32; Gowans and Campbell, *Fort Bridger*, 115; report by James F. Rusling, 11 October 1866, House Ex. Doc. 45, 39th Cong. 2d sess., serial 1289, 51; Col. J. J. Dana to Meigs, 20 November 1866, RG 92, Entry 225, Box 210, NAB; Act of 24 February 1871, maps in Drawer 189, Wyo. 7–2, Cartographic Division, National Archives at College Park. Some confusion exists about Fort Bridger's reduced size. U.S. War Department, Surgeon-General's Office, Circular No. 8, *A Report on the Hygiene of the United States Army with Descriptions of Military Posts,* 316, described the reservation as "four miles north by one mile east and west," or 4 square miles. The next year, however, it was described as "sixteen square miles," or 4 miles by 4 miles, in *Outline Descriptions of the Posts in the Military Division of the Missouri, commanded by Lieutenant General P. H. Sheridan,* 88. Gowans and Campbell, *Fort Bridger*, 126, describe it both as "four miles square" and "four square miles."

24. Robinson, *American Forts,* 176; Hazen to Adjutant General, Department of the Platte, 29 August 1866, RG 393, Records of U.S. Army Continental Commands, Part I, Entry 3731, copy in vertical files, Sheridan County Fulmer Public Library, Sheridan, Wyoming; Gwilliam, "A Study of American Army Transitional Construction Process," 181; Murray, *Military Posts in the Powder River Country*, 31, 43. Hazen's praise is repeated in Carrington, *Abasaraka, Home of the Crows,* 153.

25. Strahorn, *Fifteen Thousand Miles by Stage,* 15.

26. Explanatory note from Woodbury, 2 August 1851, RG 92, Office of the Quartermaster General, Entry 225, Box 531, NAB; drawings of block house in RG 77, Dr. 152, #8, National Archives at College Park; Gordon Chappell, "The Fortifications of Old Fort Laramie," in Murray, ed., *Fort Laramie,* 65.

27. Totten to Scott, 14 August 1850, and F. H. Masten to Jesup, 22 November 1850, RG 92, Entry 225, Box 531, NAB.

28. Cited in Mattison, "The Army Post on the Northern Plains, 1865–1895," 21; Gordon Chappell, pers. comm., 2 May 1995.

29. Chappell, "The Fortifications of Old Fort Laramie," 68–74.

30. Sherman to Cooke, 31 August 1866, RG 393, Part I, Entry 3731, Box 1, NAB; General Order 15, 5 September 1866, Department of the Platte, RG 393, Entry 3740, Box 1, F. E. Warren Air Force Base, Cheyenne, Wyoming.

31. Gowans and Campbell, *Fort Bridger*, 106.

32. Rickey, *Forty Miles a Day on Beans and Hay,* 90–106; Anderson, "Army Posts, Barracks and Quarters," 446–447.

33. Halleck, *Elements of Military Art and Science,* 105–107; Ware, *The Indian War of 1864,* 207–209; *Army and Navy Journal* 15 (6 October 1877): 138.

CHAPTER 2. HUTS AND ADOBE BUILDINGS SADLY IN NEED OF REPAIR

1. Bushman, *The Refinement of America,* 101, 242, and 247; Dell Upton, "Form and User: Style, Mode, Fashion, and the Artifact," in Pocius, ed., *Living in a Material World,* 159–160. Cary Carson, "English," in Upton, ed., *America's Architectural Roots,* 57, points out that the Georgian plan was not only English, but also reflected currents sweeping Europe in the eighteenth century.

2. "Fort Laramie Post Return, March 1867," Fort Laramie National Historic Site; "Description of Buildings at Fort Laramie, D. T.," March 1867, RG 92, Entry 225, Box 532, NAB.

3. *ARSW* (1852), 116–117.

4. King, *"Laramie;" or, the Queen of Bedlam;* Simonin, *The Rocky Mountain West in 1867,* 84.

5. John C. Kelton, "Annual Report of the Inspection of Public Buildings at Fort Laramie, W. T. June 20th, 1856," RG 92, Office of the Quartermaster General, Entry 225, Box 531, NAB.

6. "Medical History of the Post [Fort Laramie]," pp. 3–20, microfilm reel H-120, Wyoming State Archives.

7. Kelton, "Annual Report, 1856."

8. "Description of Buildings"; Dandy to Myers, 16 October 1866, RG 393, U.S. Army Continental Commands, Part I, Entry 3887, Box C–D, NAB.

9. "Description of Buildings."

10. Maj. W. W. Blunt, 10 December 1873, Fort Laramie National Historic Site.

11. U.S. War Department, Surgeon-General's Office, Circular No. 4, *A Report on Barracks and Hospitals, with Descriptions of Military Posts,* xviii, 348.

12. Simonin, *Rocky Mountain West,* 84–85; Collins to Josie, Christmas 1863, in Spring, ed., "An Army Wife Comes West," 13; plans and elevations of post trader's house, RG 92, Entry 225, Box 531, NAB; Ward to Dandy, 11 July 1866, RG 92, Entry 225, Box 531, NAB.

13. Col. J. J. Dana to Meigs, 20 November 1866, RG 92, Entry 225, Box 210, NAB.

14. Typescript of U.S. Army Commands, Letters Received, Fort Bridger, 1860–August 1868, Wyoming State Archives; Rusling to Meigs, 11 October 1866, House Exec. Doc. 45, 39th Cong., 2d sess. (serial 1289), 51.

15. U.S. War Department, Surgeon-General's Office, Circular No. 8, *A Report on the Hygiene of the United States Army with Descriptions of Military Posts,* xv–xvi, 323–324. Company B in its newly repaired barracks recorded 48 cases, or 820.8 per thousand, compared to 24 cases, or 393.4 per thousand, for Company C in its deteriorated building.

16. Surgeon-General's Office, *Report on Hygiene,* 316–317; M. E. Waters, "Medical History of the Post [Fort Bridger]," 11 and 20 September 1868, microfilm H-120, Wyoming State Archives.

17. *ARSW* (1871), 122–123; *Army and Navy Journal* 7 (11 December 1869): 258; 14 (13 January 1877): 366; and 25 (24 September 1887): 163.

18. Drew to Ludington, 25 April 1879, RG 92, Entry 225, Box 533, NAB; A. H. Appel, post surgeon, estimate for repairs and additions to post hospital, 30 June 1891, RG 92, Entry 225, Box 951, NAB.

19. Maynadier to Easton, 31 March 1866, RG 92, Entry 225, Box 532, NAB.

20. Carrington, *My Army Life,* xxii, xxiv; Marcy to Bvt. Maj. Gen. W. A. Nichols, 3 October 1867, RG 393, Part I, Entry 2546, NAB; *Army and Navy Journal* 24 (5 February 1887): 545.

21. Utley, *Frontiersmen in Blue,* 33–34; *Cheyenne Daily Leader* 21 July 1883.

22. Griess, "Dennis Hart Mahan," 180, 359–360; Ambrose, *Duty, Honor, Country,* 90, 98, 100; Forman, *West Point,* 82; Mahan, *Descriptive Geometry;* Weigley, *History of the United States Army,* 160–161.

23. *ARSW* (1870), 5, and (1883), 84–85; Senate Reports, 45th Cong., 3d sess., No. 555, 487–488, cited in Utley, *Frontier Regulars,* 83. At Fort D. A. Russell, extra-duty pay was 35

cents a day until 1884, when it increased to 50 cents. Kendall, "History of Fort Francis E. Warren," 12.

24. W. B. Hazen, 16 October 1866, House Exec. Doc. 45, 39th Cong., 2d sess. (Serial 1289); Contract, 11 September 1867, RG 92, Entry 225, Box 951, NAB; *Army and Navy Journal* 5 (4 January 1868): 313.

25. Dandy to Starring, 2 September 1866, RG 393, Part I, Entry 3887, Box C–D, NAB; Chief Quartermaster to E. E. Camp, 18 January 1867, RG 92, Entry 225, Box 531, NAB; McDermott and Sheire, "1874 Cavalry Barracks, Fort Laramie National Historic Site, Historic Structure Report, Historical Data Section," 48.

26. Meigs to Grant, 12 January 1866, RG 393, Part I, Box 2546, NAB; Hurt, "The Construction and Development of Fort Wallace, Kansas, 1865–1882," 46.

27. U.S. War Department, *Regulations for the Army of the United States, 1913,* 206.

28. Clary, "These Relics of Barbarism," 203; Dandy to Myers, 16 October 1866, RG 393, Part I, Entry 3887, Box C–D, NAB; *ARSW* (1872), 142. For size of companies, see Utley, *Frontiersmen in Blue*, 22; and Utley, *Frontier Regulars*, 16.

29. Surgeon-General's Office, *Report on Barracks and Hospitals*, xi.

30. U.S. War Department, *Regulations for the Army of the United States, 1861*, 124; *Army and Navy Journal* 7 (2 April 1870): 514. Similarly, *Army and Navy Journal* 5 (28 December 1867): 298; 6 (20 February 1869): 423; and 24 (30 April 1887): 790.

31. 31 May 1891–2 June 1891, RG 393, Part V, Entry 27, Pages 13–14, NAB; Summerhayes, *Vanished Arizona*, 9–10; *Army and Navy Journal* 24 (30 April 1887): 790. For a view in support of the custom of bumping, see *Army and Navy Journal* 27 (14 September 1889): 47.

32. U.S. War Department, *Regulations for the Army, 1861*, 124; Surgeon-General's Office, *Report on Hygiene*, 368.

33. Report of John E. Smith, 8 September 1873, McDermott files, Fort Laramie National Historic Site; Bubb to Chief Quartermaster, 25 May 1875, RG 393, Part V, Entry 28, 1:188, NAB.

34. Mackay to Jesup, 1 November 1849, RG 92, Entry 225, Box 531, NAB; Maynadier to Easton, 31 March 1866, RG 92, Entry 225, Box 532, NAB; Dandy to Myers, 16 October 1866, RG 393, Part I, Entry 3887, Box 1866–1867 C–D, NAB; Hoffman to Corley, 19 August 1856, McDermott files, Fort Laramie National Historic Site; W. B. Hazen, 16 October 1866, House Exec. Doc. 45, 39th Cong., 2d sess. (serial 1289), 7; Spring, ed., "An Army Wife Comes West," 10; Johnson, *Tending the Talking Wire,* ed. Unrau, 73, 89.

35. Hoffman to Corley, 19 August 1856, McDermott files, Fort Laramie National Historic Site; Myers to Meigs, 6 March 1867, cited in Mattes, "Surviving Army Structures at Fort Laramie National Monument," 1943, Fort Laramie National Historic Site; Commander, Fort Laramie, to Litchefield, 13 October 1866, McDermott files, Fort Laramie National Historic Site; Warrens to Chief Quartermaster, 27 June 1873, McDermott files, Fort Laramie National Historic Site.

36. Hoffman to Page, 2 June 1856, McDermott files, Fort Laramie National Historic Site; Fallett to Van Vliet, 21 January 1859, McDermott files, Fort Laramie National Historic Site; Commander, Fort Laramie, to Litchfield, 13 October 1866, McDermott files, Fort Laramie National Historic Site; Carrington, *Absaraka, Home of the Crows,* 141.

37. Records of U.S. Army Commands, Letters Received, 1860–August 1868, Fort Bridger, Wyoming State Archives; Anderson, "Army Posts," 440; Fred B. Kniffen and Henry Glassie, "Building in Wood in the Eastern United States: A Time-Place Perspective," in Upton and Vlach, eds., *Common Places,* 167.

38. *ARSW* (1871), 130, and (1869), 30; Mackay to Jesup, 1 November 1849, RG 92, Entry 225, Box 531, NAB; Hoffman to Corley, 19 August 1856, Fort Laramie National Historic Site; Spring, "An Army Wife Comes West," 10.

39. David Murphy, "Building in Clay on the Central Plains," in Carter and Herman, eds., *Perspectives in Vernacular Architecture, III,* 75–76, 78.

40. Merrill Mattes, "Preliminary Report on the Evolution of Public Buildings, Old Fort Laramie" (May 1938), 19, citing Rockafellow diary, 25 July 1865 entry, Fort Laramie National Historic Site; "Medical History of the Post [Fort Laramie]," microfilm reel H-120, p. 126, Wyoming State Archives; *Army and Navy Journal* 20 (30 September 1882): 199.

41. Hoffman to Curley, 19 August 1856, McDermott files, Fort Laramie National Historic Site; Easton to Dandy, 17 July 1866, RG 393, Part I, Entry 3887, Box E–F, 1866–1867, NAB.

42. Anderson, "Army Posts," 440.

43. *ARSW* (1874), v, cited in Pitcaithley, "The Third Fort Union," 129, 131.

44. *ARSW* (1870), 153; Meigs to Adjutant General, 23 August 1871, microfilm 619, roll 449, NAB.

45. *ARSW* (1874), v; Surgeon-General's Office, *Report on Hygiene*, 349.

CHAPTER 3. A BEAUTIFUL VILLAGE

1. Brackett, "Fort Bridger," 114; Elizabeth J. Reynolds Burt, "An Army Wife's Forty Years in the Service, 1862–1902," Elizabeth J. Burt Papers, Manuscript Division, Library of Congress, 192; Simonin, *The Rocky Mountain West in 1867,* 84; Forsyth, *The Story of the Soldier,* 110. Other village descriptions include one of Fort Dallas cited in Coffman, *The Old Army,* 116; one of Fort Union cited in Gwilliam, "A Study of American Army Transitional Construction Process during the Nineteenth Century Trans-Mississippi Frontier Period," 171; and one of Fort Lyon in Roe, *Army Letters from an Officer's Wife,* 6.

2. Coffman, *The Old Army,* 219; D. W. Meinig, "Symbolic Landscapes: Some Idealizations of American Communities," in Meinig, ed., *The Interpretation of Ordinary Landscapes,* 164–165; Reps, *The Making of Urban America,* 120, 124–125; Wood, *The New England Village,* 2; Wood, "'Build, Therefore, Your Own World,'" 32; Conforti, *Imagining New England,* 129–144; Brown, *Inventing New England,* 8, 133.

3. Andrew Jackson Downing's article, reprinted in his book *Rural Essays,* 241; Archer, "Country and City in the American Romantic Suburb," 150; Jackson, *American Space,* 33.

4. Sherman to Cooke, 11 September 1866 and 31 August 1866, RG 393, U.S. Army Continental Commands, Part I, Entry 3731, Box 1, NAB; *ARSW* (1867), 36, and (1871), 24.

5. *ARSW* (1878), 5.

6. *ARSW* (1875), 5; Utley, *Frontier Regulars,* 247; White, *"It's Your Misfortune,"* 104–105.

7. *ARSW* (1871), 36 and 24; (1877), v; (1880), vi, 4–5; and (1882), 10–14.

8. Jamieson, *Crossing the Deadly Ground,* 48–50.

9. Utley, *Frontiersmen in Blue,* 110, 346; Fellman, *Citizen Sherman,* 262; Wooster, *The Military and United States Indian Policy, 1865–1903,* 141.

10. Weigley, *History of the United States Army,* 271; Gwilliam, "A Study of American Army Transitional Construction Process during the Nineteenth Century Trans-Mississippi Frontier Period," 24.

11. Wooster, *The Military and United States Indian Policy*, 14 and 87; Weigley, *A History of the United States Army*, 271; Coffman, *The Old Army,* 245–246.

12. Weigley, *A History of the United States Army*, 158.

13. Weigley, *A History of the United States Army*, 275–281; Gwilliam, "A Study of American Army Transitional Construction Process," 25; Robert M. Utley, "The Contribution of the Frontier to the American Military Tradition," in Tate, ed., *The American Military on the Frontier,* 7–9; Wooster, *The Military and United States Indian Policy*, 56–57; Jamieson, *Crossing the Deadly Ground,* 70; Slotkin, *Fatal Environment,* 345; Coffman, *The Old Army,* 268, 400–402.

14. *ARSW* (1883), 47; (1867), 36; and (1869), 29; Guentzel, "The Department of the Platte and Western Settlement, 1866–1877," 411.

15. Dodge to Augur, 27 June 1867, RG 393, Entry 3733, copy at F. E. Warren Air Force Base. Sherman wrote, "The construction of the Union Pacific was deemed so important

that the President, at my suggestion, constituted . . . the new Department of the Platte . . . with orders to give amply protection to the working parties, and to afford every possible assistance in the construction of the road." Cited in Adams, *The Post near Cheyenne,* 3.

16. Stelter, "The Urban Frontier, 43; Kendall, "History of Fort Francis E. Warren," 6–8; Frazer, *Forts of the West,* xxiv.

17. Utley, *Frontier Regulars,* 335; Athearn, *Union Pacific Country,* 205; Adams, *The Post near Cheyenne,* 87; *ARSW* (1885), 61, 124, 141; *Army and Navy Register* 6 (12 September 1885): 587; *ARSW* (1886), 18; W. L. Carpenter, Camp Medicine Butte, to Quartermaster General, 31 March 1886, RG 393, Part I, Entry 3898, Boxes 3 and 4, NAB.

18. Stelter, "The Urban Frontier," 5.

19. Reps, *Cities of the American West,* 532; Stelter, "The Urban Frontier," 38, 43.

20. Stelter, "The Urban Frontier," 31–32, 59, 102, 262; *Chicago Tribune* cited in *Army and Navy Journal* 6 (10 April 1869): 539; Adams, *Post near Cheyenne,* 46; Burt, "An Army Wife's Forty Years," 203; Mattes, *Indians, Infants and Infantry,* 176.

21. Stelter, "The Urban Frontier," 46; Adams, *The Post near Cheyenne,* 12–13; Simonin, *The Rocky Mountain West in 1867,* 3; *Cheyenne Daily Leader,* 20 January 1883.

22. Stelter, "The Urban Frontier," 218, 283–289.

23. Stelter, "The Urban Frontier," 307–308; *Cheyenne Daily Leader,* 6 July 1883 and 20 September 1883.

24. Stelter, "The Urban Frontier," 404–405, 396, 432–436; *Cheyenne Daily Leader,* 30 December 1882; Laramie County Historical Society, *Early Cheyenne Homes, 1880–1890.*

25. Beecher, *Norwood: or, Village Life in New England;* Stowe, *Oldtown Folks;* Freeman, *A New England Nun and Other Stories;* Jewett, *Country By-Ways.*

26. U.S. War Department, Surgeon-General's Office, Circular No. 4, *A Report on Barracks and Hospitals, with Descriptions of Military Posts,* 342; Charles H. Alden, "Military History of the Post [Fort D. A. Russell]" (1869), 16, F. E. Warren Air Force Base. A possible precedent for this unusual plan was Lincoln General Hospital in Washington, D.C., built during the Civil War, in which the wards were arranged in a V-shaped plan, with twenty individual wards facing the center obliquely. Plan reproduced in Goodwin and Associates, "National Historic Context for Department of Defense Installations, 1790–1940," 2:fig. III-40; and Thompson and Goldin, *The Hospital,* 176–177.

27. McDermott and Sheire, "1874 Cavalry Barracks, Fort Laramie National Historic Site, Historic Structure Report, Historical Data Section," 32–34.

28. Sherman to Rawlins, 31 August 1866, House Exec. Doc., 39th Cong., 2d sess., vol. 6, no. 23, p. 9 (serial 1288).

29. Mattes, *Indians, Infants and Infantry,* 229.

30. Anderson, "Army Posts," 445; *Army and Navy Journal* 24 (9 April 1877): 730.

31. Coffman, *The Old Army,* 78–81, 391; Surgeon-General's Office, *Report on Barracks and Hospitals,* 343; U.S. War Department, Surgeon-General's Office, Circular No. 8, *A Report on the Hygiene of the United States Army with Descriptions of Military Posts,* 369; *Army and Navy Review* 7 (3 April 1886): 220. Although designed in 1884, the administration building was not built until 1885. The L-shaped building was reversed in plan, so that the chapel ell was on the other side than that indicated in the 1884 drawing.

32. Clary, "These Relics of Barbarism," 119; Surgeon-General's Office, *A Report on Barracks and Hospitals,* 349; E. H. Brooke to Quartermaster General, 18 May 1889, RG 393, Part V, Entry 28, 5:268, NAB.

33. "Medical History of the Post [Fort Bridger]," microfilm reel H-120, Wyoming State Archives; Surgeon-General's Office, *Report on Barracks and Hospitals,* 341; James Regan to Chief Quartermaster, 28 March 1885, RG 92, Entry 225, Box 952, NAB; Received from Chief Quartermaster, 22 May 1904, RG 393, Part V, Entry 27, 456, NAB. The "cult of the tree" is discussed by Rees, *New and Naked Land,* 95.

34. Burt, "An Army Wife's Forty Years," 192–193; Carrington, *Absaraka, Home of the Crows,* 106. Headquarters of the Division of the Missouri sent four mowing machines to

Fort Reno in 1866. Easton to Myers, 16 July 1866, RG 393, Part I, Entry 3887, Box 1866–1867 E–F, NAB.

35. Downing, *Rural Essays*, 242; Carrington, *Absaraka, Home of the Crows*, 148; Acting Assistant Quartermaster to Chief Quartermaster, 8 April 1881, RG 393, Part V, Entry 28, 2:79, NAB.

36. Cited in Hafen and Young, *Fort Laramie*, 168; Surgeon-General's Office, *Report on Barracks and Hospitals*, 349, 360; "Medical History of the Post [Fort Laramie]," entry for 4 October 1869, microfilm reel H-120, Wyoming State Archives.

37. Forsyth, *The Story of the Soldier*, 113; *Army and Navy Journal* 10 (14 June 1873): 702.

38. *Army and Navy Journal* 29 (23 January 1892): 383; From Robinson, 3 May 1881, RG 393, Part V, Entry 29, 1:42, NAB; cited in Hedren, *Fort Laramie in 1876,* 40.

39. Risch, *Quartermaster Support of the Army, 1775–1939,* 490; Richard Harding Davis in *Harpers Weekly*, cited in *Army and Navy Journal* 29 (25 June 1892): 772.

40. U.S. manuscript censuses for Fort Laramie, 1860 and 1880, transcriptions at Fort Laramie National Historic Site; recruiting information from *ARSW* (1883), 84–85; information on African Americans from Coffman, *The Old Army*, 365. Other historians have found that these percentages of foreign-born soldiers were consistent with nationwide trends. Coffman, *The Old Army*, 141, 330; Utley, *Frontiersmen in Blue*, 40; Utley, *Frontier Regulars*, 23.

41. 1880 census, Fort Laramie; Foner, *The United States Soldier between Two Wars,* 15; Murray, "Prices and Wages at Fort Laramie, 1881–1885," 19.

42. *Cheyenne Daily Leader*, 2 August 1883, 3; *ARSW* (1875), 5; General Order 36, Fort D. A. Russell, 10 December 1868, F. E. Warren Air Force Base; Stewart, "Army Laundresses," 431–434; *Army and Navy Journal* 20 (23 June 1883): 1054. After the army discontinued its support of laundresses, enlisted men's wives continued to wash soldiers' laundry at rates set by the post council of administration.

43. *ARSW* (1875), 6–7; Foner, *The United States Soldier*, 13; Stallard, *Glittering Misery,* 53–54; General Order 36, Fort D. A. Russell, 10 December 1868, F. E. Warren Air Force Base.

44. Myres, "Romance and Reality on the American Frontier," 409–427, especially n. 4; Carrington, *My Army Life;* Custer, *"Boots and Saddles";* Jeffrey, *Frontier Women*.

45. Frances Carrington, *My Army Life*, 61; Margaret Carrington, *Absaraka*, dedication. Frances was the wife of George Grummond, a lieutenant who was killed in the Fetterman Massacre. After Henry Carrington's first wife, Margaret, died in 1870, he married Frances.

46. *Army and Navy Journal* 20 (14 April 1883): 835; and 24 (30 April 1887): 790.

47. *Army and Navy Journal* 20 (7 April 1883): 811.

48. Burt, "An Army Wife's Forty Years," 90. Recent studies of officers' wives include Stallard, *Glittering Misery;* Myres, "Frontier Historians, Women, and the 'New' Military History," 27–37; and Leckie, "Reading between the Lines," 137–160.

49. Burt, "An Army Wife's Forty Years," 90; 1870 census, Fort Laramie. Ten years later, the situation was similar, with twelve officers, ten with wives, according to the 1880 census, Fort Laramie. The 1860 census indicated no women at the post, other than laundresses.

50. 1870 and 1880 censuses, Fort Laramie; Hedren, *Fort Laramie in 1876*, 41.

51. 1870 and 1880 censuses, Fort Laramie; Rusling to Meigs, 11 October 1866, House Exec. Doc. 45, 39th Cong., 2d sess. (Serial 1289), 53; Custer, *"Boots and Saddles,"* 98.

52. General Order 58, Department of the Platte, 11 September 1871, F. E. Warren Air Force Base; 1860 census, Fort Laramie.

53. Carter to O. D. Filley, no date (probably 1862), Carter Collection, American Heritage Center; Rusling to Meigs, 11 October 1866, House Exec. Doc. 45, 39th Cong., 2d sess. (Serial 1289), 52; Jno W. Bubb, 25 September 1874, RG 393, Part V, Entry 28, 1:73, NAB; Received from Chief Quartermaster, 24 May 1878, RG 393, Part V, Entry 29, 1:153, NAB:

Carter, "Fort Bridger in the Seventies," 112–113; Rusling, *Across America,* 158; M. E. Carter to General J. R. Brooke, 5 August 1890, RG 92, Entry 225, Box 209, NAB; *Cheyenne Daily Leader,* 22 May 1895.

54. Merrill J. Mattes, "The Sutler's Store," in *Fort Laramie,* 36–37; Spring, "Old Letter Book Discloses Economic History of Fort Laramie, 1858–1871," 242; Ward to Dandy, 11 July 1866, and Dandy to Easton, 11 July 1866, RG 92, Entry 225, Box 531, NAB; *Cheyenne Daily Leader,* 8 March 1876 and 22 March 1876, transcribed WPA files subject 98, Wyoming State Parks and Historic Sites Division.

55. W. Cox, post adjutant, to W. H. Brown, 27 October 1868, and W. McCammon, post adjutant, to W. Brown, 16 July 1871, RG 98, transcribed in McDermott files, Fort Laramie National Historic Site; Coffman, *The Old Army,* 359–360; Foner, *The United States Soldier,* 92–94.

56. Carrington, *Absaraka,* 76–77; *Army and Navy Journal* 3 (11 August 1866): 806; Mattes, *Indians, Infants and Infantry,* 157.

57. Surgeon-General's Office, *Report on Barracks and Hospitals,* 349–350; Mattes, *Indians, Infants and Infantry,* 27, 175.

CHAPTER 4. THE HOMELIKE APPEARANCE OF THE HOUSE

1. Summerhayes, *Vanished Arizona,* 12; U.S. War Department, Surgeon-General's Office, Circular No. 4, *A Report on Barracks and Hospitals, with Descriptions of Military Posts,* 342.

2. Summerhayes, *Vanished Arizona,* 17; Elizabeth J. Burt, "An Army Wife's Forty Years in the Service, 1862–1902," Elizabeth J. Burt papers, Manuscript Division, Library of Congress, 200; Proceedings of Board of Officers ordered 31 March 1870, by Col. John H. King, RG 92, Office of the Quartermaster General, Entry 225, Box 953, NAB.

3. King, *Captain Blake,* 141; Charles H. Alden, "Medical History of the Post [Fort D. A. Russell]," F. E. Warren Air Force Base, 17; Proceedings of a Board of Officers ordered 31 March 1870, RG 92, Entry 225, Box 953, NAB; Mattes, *Indians, Infants and Infantry,* 179; Surgeon-General's Office, *Report on Barracks and Hospitals,* 342. The barracks had the same construction as the officers' quarters, except that the walls were lined with adobe bricks for added insulation. The roofs of the officers' quarters were replaced with sheet iron in 1871, according to "Medical History of the Post [Fort D. A. Russell]," December 1871 entry.

4. Surgeon-General's Office, *Report on Barracks and Hospitals,* 342.

5. Surgeon-General's Office, *Report on Barracks and Hospitals,* 341.

6. U.S. War Department, Surgeon-General's Office, Circular No. 8, *A Report on the Hygiene of the United States Army, with Descriptions of Military Posts,* 368.

7. Regan to Quartermaster General, 19 April 1886, RG 92, Entry 225, Box 953, NAB.

8. Grashof, "Standardized Plans, 1866–1940," 2:OQ-41; Agreement, Humphrey and John A. Rishel and E. Russell Ellis, 8 November 1887, RG 92, Entry 225, Box 951, NAB. By June 1888 Rishel and Ellis had defaulted on their contract, and Humphrey was forced to readvertise. One of the 1885 quarters on the original parade ground survives, known today as Building No. 1. Of the four built off the parade ground, three survive as Buildings Nos. 22, 24, and 25. Three of the 1888 quarters survive as Buildings Nos. 26, 27, and 28.

9. Glassie, *Folk Housing in Middle Virginia,* 37; Carter, "Traditional Design in an Industrial Age," 424, 437.

10. Regan to Quartermaster General, 26 April 1884, RG 92, Entry 225, Box 953, NAB; King, *Captain Blake,* 141–143; Regan to Quartermaster General, 19 April 1886, RG 92, Entry 225, Box 953, NAB; Report of Captain Humphrey, October 1888, RG 92, Entry 225, Box 953, NAB. The frame officers' quarters probably also received bathrooms in the early 1890s, but no documentation of this could be found.

11. *ARSW* (1882), 13; and (1883), 126.

12. Stivers to Quartermaster General, 23 April 1884, RG 393, Part V, Entry 28, 3:2–15, NAB.

13. Burt, "An Army Wife's Forty Years in the Service 1862–1902," 88–89; photograph of Carter in front of his remodeled house, indicating that the remodeling occurred before his death in 1881, reproduced in Ellison, *Fort Bridger, Wyoming,* 66.

14. Received from Dandy, 18 February 1887, RG 393, Part V, Entry 29, 3:347, NAB.

15. C. St. J. Chubb, "List of Government Buildings at Fort Bridger, Wyoming" (6 November 1890), RG 92, Entry 225, Box 209, NAB.

16. *Omaha Herald,* 20 October 1874, from McDermott files, Fort Laramie National Historic Site.

17. U.S. War Department, *Regulations for the Army, 1861,* 125; Anderson, "Army Posts, Barracks and Quarters," 445, 435; *Army and Navy Journal* 24 (30 April 1887): 790.

18. Frank A. Page to Bvt. Maj. H. Gardner, 7 November 1867, Letters Received by Office of Adjutant General (main series) 1861–1870, microcopy no. 619, roll 449, NAB; Report of the Marcy Board, 42d Cong., 3d sess., House Report No. 85, 1 March 1873, 147.

19. Meigs to Secretary of War, 17 February 1875; Marcy to Secretary of War, 25 February 1875; Meigs to Secretary of War, 5 March 1875, all in RG 92, Entry 225, Box 532, NAB. Marcy to Secretary of War, 14 and 24 April 1875, RG 92, Entry 225, Box 952, NAB.

20. *Army and Navy Journal* 20 (7 April 1883): 811.

21. Fisch and Wright, *The Story of the Noncommissioned Officer Corps,* 72; *ARSW* (1881), 34; James Regan to Quartermaster General, 19 April 1886, RG 393, Part I, Entry 3898, Box 3, NAB; Estimate of material, 29 April 1886, RG 393, Part I, Entry 3898, Box 15, NAB.

22. The 1884 drawings indicate a building of about 220 feet in length, while the HABS drawings of the ruins describe it as 242 feet in length. The width, 27 feet 6 inches, was the same on both. Sheire, "Historic Structure Report, Part I, Non-Commissioned Staff Officers' Quarters, Fort Laramie National Historic Site, Wyoming," 3; Louis Brechemin, interviewed by David L. Hieb, 24 July–4 August 1948, McDermott files, Fort Laramie National Historic Site.

23. *ARSW* (1881), 36; (1882), 12; and (1880), 5.

24. *Army and Navy Journal* 20 (28 April 1883): 879; 25 (14 April 1883): 836 (emphasis in original).

25. *Army and Navy Journal* 20 (14 April 1883): 835; 48th Cong., 1st sess., House Report No. 1433 (serial 2257), 3; *Army and Navy Review* 5 (21 June 1884): 5; and 4 (5 July 1884): 1.

26. Contract between Col. Carling, Fort D. A. Russell, and Mason & Co. to provide lumber, 2 September 1867, Records of Contracts August 1866–November 1869, RG 393, Entry 3893. Also see Col. Carling to Bvt. Maj. Gen. C. C. Augur, 21 November 1867: "If you can send me 100,000 feet of boards, 16 feet long, within a week, should like them" (RG 393, Entry 3733, F. E. Warren Air Force Base).

27. *ARSW* (1882), 12; and (1883), 128.

28. Woodbury to Totten, 2 August 1851, Fort Laramie National Historic Site; *ARSW* (1869), 243–244.

29. *Cheyenne Daily Leader,* 10 November 1882; Humphrey to Chief Quartermaster, 9 June 1888, F. E. Warren Air Force Base.

30. *AR SW* (1858), 797.

31. This experiment is described more fully in Nowak, "From Fort Pierre to Fort Randall," 95–116.

32. Donaldson to Myers, 1 June 1867, RG 393, Part I, Entry 3887, Box C–D, NAB; Easton to Myers, 5 July 1867, RG 393, Part I, Entry 3887, Box E–F, NAB.

33. Burt, "An Army Wife's Forty Years," 200; Surgeon-General's Office, *Report on Barracks and Hospitals,* 342.

34. Stelter, "The Urban Frontier," 37–38.

35. "Historic Structures Report: Stabilization and Rehabilitation of the Old Army Bridge," 1962, Fort Laramie National Historic Site.

36. Quartermaster General's Office, *Notes on Building in Concrete*, 9; Surgeon-General's Office, *Report on Barracks and Hospitals*, 341–342.

37. Fowler, *The Octagon House*, viii, xi, 18; Wheaton, "Lime-Grout and Fort Laramie, Wyoming," 2; O'Dell, "Building Technology in the Department of the Platte, 1866–1890," 51.

38. R. M. O'Reilly to Surgeon General, 22 November 1872, in "Medical History of the Post [Fort Laramie]," microfilm reel H-120, 270, Wyoming State Archives; endorsement by Sheridan, 20 December 1872, RG 92, Entry 225, Box 532, NAB; "Survey Report for Stabilization Measures to Hospital Ruins, Building No. 13," 6–7.

39. Warrens to Quartermaster General, 27 June 1873, RG 92, Entry 225, Box 532, NAB; "Medical History of the Post [Fort Laramie]," 31 August 1874 entry, microfilm reel H-120, Wyoming State Archives.

40. Cuny and Ecoffey apparently obtained government contracts. A contract between the chief quartermaster in Omaha and Jules Ecoffey to supply 1,200 cords of wood to Fort Laramie is referred to in Grimes to Myers, 24 April 1867, RG 393, Part I, Entry 3898, Box 15, NAB.

41. O'Dell, "Building Technology," 60.

42. Anderson, "Army Posts," 441–442.

43. Proceedings of a Board of Officers convened at Fort Robinson, Nebraska, 16 April 1886, RG 393, Part I, Entry 3898, Box 3, NAB; McParlin, first endorsement, Ogle to Medical Director, 17 February 1888, RG 393, Part V, Entry 29, 3:434, NAB.

44. Sherman, third endorsement to letter from J. M. Robertson to Townsend, 7 August 1871, and Meigs to Adjutant General, 23 August 1871, microfilm 619, roll 449, NAB; *ARSW* (1872), 280.

45. Custer, *"Boots and Saddles,"* 174.

46. Anderson, "Army Posts, Barracks and Quarters," 444, 434.

47. Fort Hartsuff's concrete guard house also appears to utilize the same plans. See photograph in O'Dell, "Building Technology," 61.

48. Sheire, "The 1876 Old Bakery, Fort Laramie National Historic Site, Historic Structure Report, Part II," 22; Bell, *Notes on Bread Making, Permanent and Field Ovens, and Bake Houses,* 87–91. A drawing labeled "proposed bake house at Fort Laramie" is very close to Bell's plan, drawing at Fort Laramie National Historic Site.

49. Starr, *The Social Transformation of American Medicine,* 150; Surgeon-General's Office, *Report on Hygiene*, viii, 324.

50. Surgeon-General's Office, *Report on Barracks and Hospitals*, 348.

51. Surgeon-General's Office, *Report on Barracks and Hospitals*, 361–362.

52. Surgeon-General's Office, *Report on Barracks and Hospitals*, xxv.

53. Billings, *The Principles of Ventilation and Heating and Their Practical Application;* Billings, *Ventilation and Heating;* Allen and Walker, *Heating and Ventilation,* 214; *Hospital Plans;* Ashburn, *A History of the Medical Department of the United States Army,* 388–389; Duffy, *The Sanitarians,* 167; Hume, "John Shaw Billings as an Army Medical Officer," 242; Reed, *The New York Public Library,* 7–8; Clary, "These Relics of Barbarism," 124.

54. Hume, "John Shaw Billings," 225–270.

55. Klauss, Tull, Roots, and Pfafflin, "History of the Changing Concepts in Ventilation Requirements," 52–53; Surgeon-General's Office, *Report on Barracks and Hospitals*, vi; Billings, *Heating and Ventilation,* 19.

56. Surgeon-General's Office, *Report on Barracks and Hospitals*, xvi, xi, xxii, xiii–xiv.

57. Bruegmann, "Architecture of the Hospital, 1770–1870," 40–43; Surgeon General's Office, Circular No. 4, "Plan for a Post Hospital of Twenty-Four Beds" (27 April 1867), RG 112, Office of the Surgeon General, Entry 63, Box 4, NAB.

58. Surgeon-General's Office, *Report on Barracks and Hospitals*, 342–343.

59. Surgeon-General's Office, *Report on Hygiene*, liv; General Orders No. 118, RG 112, Entry 63, Box 3, NAB; reprinted in *Army and Navy Journal* 8 (3 December 1870): 246; Ogle to Medical Director, 17 February 1888, RG 393, Part V, Entry 28, 4:431, NAB. The contemplated hospital was never approved for construction.

60. Surgeon General's Office, Circular No. 3 (23 November 1870), RG 112, Entry 63, Box 3, NAB; Surgeon-General's Office, *Report on Barracks and Hospitals*, xxii; Surgeon General's Office, Circular No. 2 (27 July 1871), RG 112, Entry 63, Box 3, NAB.

61. "Medical History of the Post [Fort Laramie]," 22 November 1872, 270, microfilm reel H-120, Wyoming State Archives.

62. C. H. Warrens to Post Adjutant, 31 March 1874, RG 92, Entry 225, Box 532, NAB; O'Reilly to Jos. B. Brown, 21 August 1873, RG 92, Entry 225, Box 532, NAB.

63. Bruegmann, "Architecture of the Hospital," 113–116.

64. Surgeon General's Office, Circular No. 10 (20 October 1877), RG 112, Entry 63, Box 1, NAB; Received from Chief Quartermaster, 21 May 1891, and Received from Post Surgeon, 3 December 1891, RG 393, Part V, Entry 27, 9, NAB.

65. Eltonhead to Chief Quartermaster, 18 May 1887, RG 393, Part V, Entry 38, 4:260; Eltonhead to Chamberlain and Small, Evanston, and George East, Cheyenne, 9 June 1887, RG 393, Part V, Entry 28, 4:281, NAB.

66. Eltonhead to Chief Quartermaster, 27 April 1887, RG 393, Part V, Entry 28, 4:238, NAB; Eltonhead to Jerrett, 23 May 1887, RG 393, Part V, Entry 28, 4:264, NAB. Jerrett, or "Jewett" as he was called in an article, was also the principal contractor for the construction of Fort Niobrara, Nebraska, which was built in 1886, mostly of adobe. *Army and Navy Review* 7 (24 April 1886): 263. Eltonhead had other problems as well. He was court-martialed in February 1886, charged with "conduct to the prejudice of good order and military discipline," for having punished a musician in his regiment by tying his hands together and dragging him behind a horse. Eltonhead was also charged with being drunk on duty during the incident. The court-martial acquitted him of both counts. *Army and Navy Review* 7 (27 March 1886): 195.

67. Humphrey to Chief Quartermaster, 20 March 1888, RG 92, Entry 225, Box 953, NAB (emphasis in original).

68. *Army and Navy Journal* 24 (21 May 1887): 849; 2 September 1883, RG 393, Part V, Entry 27, 1:368, NAB.

69. *Army and Navy Journal* 24 (30 April 1887): 790; and 24 (21 May 1887): 849. For more on Shoppell, see Garvin, "Mail-Order House Plans and American Victorian Architecture," 314–321.

CHAPTER 5. QUITE AN AIR OF HOME COMFORT

1. *Army and Navy Journal* 15 (9 February 1878): 419. The officers' wives were the spouses of a lieutenant, a major, and a colonel; the hostess was the wife of a general.

2. Dubois, "A Social History of Cheyenne, Wyoming, 1875–1885," 1–2, 14, 15, 19, 31.

3. *ARSW* (1883), 47; and (1887), 133.

4. *Army and Navy Journal* 5 (11 January 1868): 330; Fitzgerald, *An Army Doctor's Wife on the Frontier*, 327.

5. Summerhayes, *Vanished Arizona*, 13; Elizabeth J. Reynolds Burt, "An Army Wife's Forty Years in the Service, 1862–1902," Elizabeth J. Burt papers, Manuscript Division, Library of Congress, 90–91; Carrington, *My Army Life*, 54.

6. Foner, *The United States Soldier between Two Wars*, 63; *ARSW* (1881), 35; Summerhayes, *Vanished Arizona*, 14–16.

7. Crossman to Canby, 4 January 1860, Letters Received 1860–August 1868, Fort Bridger, Wyoming State Archives; Stivers to Chief Quartermaster, 1 June 1884, RG 393, U.S. Army Continental Commands, Part V, Entry 28, 3:52, NAB; Truitt to Chief Quartermaster, 21 November 1885, RG 393, Part V, Entry 28, 3:257, NAB; Ogle to Henry Lehmann,

Omaha, 12 December 1887, RG 393, Part V, Entry 28, 4:425, NAB; Eltonhead to McCanley, 28 December 1886, RG 393, Part V, Entry 28, 4:149, NAB.

8. *Army and Navy Journal* 5 (5 October 1867): 106; *Army and Navy Journal* (22 July 1867), cited in Foner, *The United States Soldier*, 18; U.S. War Department, Surgeon-General's Office, Circular No. 4, *A Report on Barracks and Hospitals, with Descriptions of Military Posts*, xvi. See also Clary, "These Relics of Barbarism"; Clary, "A Life Which Is Gregarious in the Extreme"; Brown, "A Pictorial History of Enlisted Men's Barracks of the U.S. Army, 1861–1895"; Brown, *The Army Called It Home*.

9. *ARSW* (1858), 797; Rickey and Sheire, "The Cavalry Barracks," 28–29; Risch, *Quartermaster Support of the Army, 1775–1939*, 488; *ARSW* (1872), 142; U.S. War Department, Surgeon-General's Office, Circular No. 8, *A Report on the Hygiene of the United States Army, with Descriptions of Military Posts*, 317.

10. *ARSW* (1878), 324; and (1885), 604; Clary, "These Relics of Barbarism," 142.

11. Rickey and Sheire, "The Cavalry Barracks," 22; Bourke, *On the Border with Crook*, 40; *Army and Navy Journal* 5 (5 October 1867): 106.

12. *ARSW* (1878), 325, 262; (1880), 333; and (1883), 142. Two years later yet another change was made: hickory replaced white oak. *ARSW* (1885), 605.

13. Surgeon-General's Office, *Report on Hygiene*, xi; Surgeon-General's Office, *Report on Barracks and Hospitals*, xiii.

14. Surgeon-General's Office, *Report on Hygiene*, ix, 355.

15. Surgeon-General's Office, *Report on Barracks and Hospitals*, 347, 342.

16. Surgeon-General's Office, *Report on Hygiene*, 348–349, 321.

17. Anderson, "Army Posts, Barracks and Quarters," 432–433.

18. Surgeon-General's Office, *Report on Barracks and Hospitals*, 361, xii; Surgeon-General's Office, *Report on Hygiene*, 317; "Medical History of the Post [Fort Bridger]," 31 December 1873, microfilm reel H-120, Wyoming State Archives. Kilns to fire the necessary bricks were built early in the construction program; Fort Laramie had brick kilns by 1851. Woodbury to Totten, 2 August 1851, Fort Laramie National Historic Site.

19. Rickey and Sheire, "The Cavalry Barracks," 23; Clary, "A Life Which Is Gregarious," 617; U.S. War Department, *Regulations for the Army of the United States, 1861*, 125; U.S. War Department, *Regulations for the Army of the United States* (1889), 115.

20. Clary, "The Role of the Army Surgeon in the West," 53–66; Ogle, *All the Modern Conveniences*, 126–128; Thomas J. Schlereth, "Conduits and Conduct: Home Utilities in Victorian America, 1876–1915," in Foy and Schlereth, eds., *American Home Life, 1880–1930*, 234–236; Handlin, *The American Home*, 460–463.

21. "Medical History of the Post [Fort Laramie]" (1868), microfilm reel H-120, Wyoming State Archives; Surgeon-General's Office, *Report on Barracks and Hospitals*, 361; Acting Inspector General to Acting Adjutant General, 25 August 1887, F. E. Warren Air Force Base.

22. Surgeon-General's Office, *Report on Barracks and Hospitals*, xvii; "Medical History of the Post [Fort Laramie]," January 1885 and June 1886, microfilm reel H-120, Wyoming State Archives.

23. Surgeon-General's Office, *Report on Barracks and Hospitals*, 349, 362, 343; Roe, *Army Letters from an Officer's Wife*, 276; Mattes, *Indians, Infants and Infantry*, 91; J. H. Taylor to O. O. Howard, 12 August 1883, cited in Foner, *The United States Soldier*, 18.

24. Perry to Sheridan, 2 May 1873, RG 92, Office of the Quartermaster General, Entry 225, Box 953, NAB; *ARSW* (1873), 9; Synopsis by Post Quartermaster James M. Moore, attached to Sheridan to Belknap, 23 April 1874, RG 92, Entry 225, Box 952, NAB; Surgeon-General's Office, *Report on Hygiene*, 268.

25. Kendall, "History of Fort Francis E. Warren," 36–38.

26. Received from Chief Quartermaster, 10 January 1885, RG 393, Part V, Entry 27, 406, NAB; Commanding officer to Assistant Adjutant General, 23 July 1885, RG 393, Part V, Entry 27, 415, NAB; Kendall, "History of Fort Francis E. Warren," 39; Adams, *The Post near Cheyenne*, 86; Regan memo, 11 May 1885, RG 92, Entry 225, Box 952, NAB.

27. "Fort Bridger Water System Reseach Report," no date, Wyoming State Parks and Historic Sites Division, 1; Eltonhead to Chief Quartermaster, 3 August 1886, RG 393, Part V, Entry 28, 3:436, NAB; Eltonhead to Chief Quartermaster, 2 November 1886, RG 393, Part V, Entry 28, 4:101, NAB; Received from Dandy, 26 July 1887, RG 393, Part V, Entry 29, 3:381, NAB; Eltonhead to Chief Quartermaster, 1 October 1887, RG 393, Part V, Entry 28, 4:368, NAB; Received from Hughes, 23 October 1887, RG 393, Part V, Entry 29, 3:396, NAB.

28. *Army and Navy Review* 9 (14 July 1888): 436; and 9 (11 August 1888): 500; Brooke to Chief Quartermaster, 27 May 1889, RG 393, Part V, Entry 28, 5:276, NAB; Received from Acting Assistant Quartermaster, Fort Duchesne, 7 November 1888, RG 393, Part V, Entry 29, 2:78, NAB.

29. "Fort Bridger Water System."

30. "Fort Laramie, Wyo. Terr., Office Brief as to Water Supply" [1887?], RG 92, Entry 225, Box 532, NAB.

31. *Army and Navy Journal* 20 (27 January 1883): 577. Aversion to cesspools is also discussed in Ogle, *All the Modern Conveniences*, 47–48. Report of Acting Inspector General, 7 November 1885, McDermott files, Fort Laramie National Historic Site.

32. Chynoweth to Chief Quartermaster, 30 January 1888, and Carey to Holabird, 25 May 1889, RG 92, Entry 225, Box 951, NAB.

33. Chynoweth to Chief Quartermaster, 7 July 1890, RG 92, Entry 225, Box 951, NAB.

34. Lieutenant Sage, 1890, RG 92, Entry 225, Box 521, NAB; *Army and Navy Journal* 24 (21 May 1887): 849.

35. Holabird to Chief Quartermaster, 3 January 1890, RG 92, Entry 225, Box 951, NAB.

36. Surgeon-General's Office, *Report on Barracks and Hospitals*, 349, xvi–xvii; Risch, *Quartermaster Support of the Army*, 488–489.

37. Surgeon-General's Office, *Report on Hygiene*, x; Clary, "A Life Which Is Gregarious," 88.

38. Two of the bathhouse/reading rooms were frame, measuring 30 by 15 feet and 36 by 15 feet, and one was frame and adobe, measuring 55 by 16 feet. Lt. E. E. Hardin, "Report of Condition, capacity, etc. of public buildings at Fort Laramie on 31st day of March 1883," RG 92, Entry 225, Box 533, NAB. Bathhouses combined with libraries were also a feature of company towns in the Michigan copper-mining district. Lankton, *Cradle to Grave*, 172.

39. Simpson to Post Adjutant, 15 April 1881, RG 92, Entry 225, Box 4333, NAB; Received from Chief Quartermaster, 10 October 1880, RG 393, Part V, Entry 27, 297, NAB; Acting Inspector General to Assistant Adjutant General, 22 November 1888, F. E. Warren Air Force Base; Acting Inspector General to Assistant Adjutant General, 30 June 1889, F. E. Warren Air Force Base.

40. Regan to Quartermaster General, 19 April 1886, RG 393, Part I, Entry 3898, Box 3, NAB; Hardin to Quartermaster General, 23 June 1884, RG 92, Entry 225, Box 533, NAB.

41. Burt, "An Army Wife's Forty Years," 91, cited in Rickey and Sheire, "The Cavalry Barracks," 25.

42. *ARSW* (1881), 13, 12, 225; and (1882), 257; *Army and Navy Journal* 18 (30 April 1881): 816.

43. *ARSW* (1881), 13; poem cited in Foner, *The United States Soldier*, 78; *Army and Navy Journal* 18 (30 April 1881): 816; Clary, "A Life Which Is Gregarious," 51.

44. Clary, "A Life Which Is Gregarious," 48–51.

45. Stelter, "The Urban Frontier," 415, 417–419, 423–431.

CHAPTER 6. A MONUMENT TO THE PORK BARREL

1. *ARSW* (1890), 5; (1882), 15–17; and (1890), 122.

2. Connolly, "Senator Warren of Wyoming," 10–11, 30; Larson, *History of Wyoming*, 180, 381, 383. Robert LaFollette, the Progressive senator from Wisconsin, repeated many of Connolly's charges, calling Warren "the 'boss' of Wyoming" who "exercised an almost dictatorial

control over the U.S. army," in an article in his weekly magazine, which was quoted and summarized on the front page of the *Cheyenne Sunday State Leader*, 15 September 1912. Although the House Committee on Expenditures in the Interior Department concluded that Warren violated the law, he apparently paid no penalty.

3. Hansen, "The Congressional Career of Senator Francis E. Warren from 1890 to 1902," 4–8; Connolly, "Senator Warren," 10–11.

4. Hansen, "The Congressional Career," 12–16; Robert C. Byrd, *The Senate, 1789–1989: Historical Statistics, 1789–1982,* ed. Wendy Wolf, 100th Cong., 1st sess., Sen. Doc. 100-20, 524, 530, 537, 554, 560. Warren chaired the following committees: Irrigation and Reclamation, 1895–1899; Claims, 1899–1905; Military Affairs, 1905–1911; Engrossed Bills, 1913–1919; Appropriations, 1911–1913, 1919–1929.

5. Connolly, "Senator Warren," 10.

6. Sherman also founded the Military Service Institution of the United States in 1878 to develop the professional identity of his officers and to unify them in the face of congressional opposition to the army in general. The institution's journal, which began publication in 1879, provided a forum for officers to debate army-related issues. Sherman also initiated a system of post–West Point graduate schools for officers, such as the School of Application for Infantry and Cavalry established at Fort Leavenworth in 1881. Skowronek, *Building a New American State,* 92. Under John M. Schofield's reign as commanding general, from 1888 to 1895, reforms continued. Schofield instituted two measures clearly aimed at professionalizing his officers: examinations on professional subjects before promotion and annual efficiency reports. Where previously, promotion had been possible only within each regiment, Schofield opened promotions to vacancies within each branch, thus improving the odds of advancement. Nenninger, *The Leavenworth Schools and the Old Army,* 34.

7. Skowronek, *Building the New American State,* 112, 213–221; Greene, "The United States Army," 606; Nenninger, *The Leavenworth Schools,* 52–53; Weigley, *History of the United States Army,* 314–326.

8. *ARSW* (1901), 8; and (1902), 21.

9. "Results of Preliminary Examination and Surveys of Sites for Military Posts," House Doc. 618, 57th Cong., 1st sess. (serial 4379), 396; *ARSW* (1906), 49; *The Nation* 83 (12 July 1906): 27.

10. *ARSW* (1911), 14, 337; and (1912), 311–312.

11. Utley, *Frontier Regulars,* 403–407; Clyde A. Milner II, "National Initiatives," in Milner, O'Connor, and Sandweiss, eds., *The Oxford History of the American West,* 182.

12. Adams, *The Post near Cheyenne,* 121; *ARSW* (1895), 4.

13. Gould, "Francis E. Warren and the Johnson County War," 135; Larson, *A History of Wyoming,* 268–284; *ARSW* (1892), 124.

14. *ARSW* (1894), 57; Larson, *A History of Wyoming,* 297–298; Kendall, "History of Fort Francis E. Warren," 27–28; Adams, *The Post near Cheyenne,* 120.

15. Adams, *The Post near Cheyenne,* 122–123.

16. *The Nation* 93 (21 December 1911): 596.

17. Wheaton, "The Architecture of the Army Post," 11. Oddly, Fort Missoula, Montana, also received buildings in the Spanish colonial style.

18. Black & Clark of Cheyenne was awarded the contract for construction at $6,610 in the fall of 1893, and construction was completed the following year. O. F. Rhoades of Cheyenne won the contract for the plumbing, and W. G. Higgins installed the steam heating. Received from Chief Quartermaster, 7 October 1893, RG 393, U.S. Army Continental Commands, Part V, Entry 27, 242, NAB; Received from W. G. Higgins, 22 April 1894, RG 393, Part V, Entry 27, 294, NAB.

19. Received from Quartermaster General, 28 September 1899, RG 393, Part V, Entry 27, 158, NAB; Received from F. J. Grodavent, 20 January 1900, RG 393, Part V, Entry 27, 168, NAB. Grodavent's connection to Fort Logan made by Grashof, "Standardized Plans,

1866–1940," 1:36, 2:3–162; and by Jennings, "Frank J. Grodavent," 1–23. Jennings (19–20) notes that Grodavent held the position of superintendent at Fort D. A. Russell in 1899–1902, 1904–1913, and 1918–1921. In 1902–1904, William DuBois, noted Cheyenne architect, filled the position. One of the wood-frame double officers' quarters removed from the post is said to exist at 2523 Evans in Cheyenne. If so, the doors and windows have been changed and new siding has been added.

20. The constructing quartermaster thanked the post quartermaster for "suggestions re lack of suitable arrangement" in five different standardized plans. Received from Constructing Quartermaster, 26 April 1904, RG 393, Part V, Entry 29, 3, NAB. Grashof, "Standardized Plans," 1:29, found little discussion of the change to standardized plans issued from Washington. The annual reports of the quartermaster general and secretary of war, the *Army and Navy Journal*, and even Erna Risch's administrative history of the Quartermaster's Department are silent on this matter.

21. Most of these "permanent" forts were built with little clear purpose. Fort William Henry Harrison may have the distinction of the shortest life; garrisoned in 1895, it was abandoned in 1913. Similarly, the army garrisoned Fort Mackenzie in 1899 and abandoned it in 1918. Of the forts mentioned, only Fort Crook is still a military base; it is now Offutt AFB. Forts Mackenzie and W. H. Harrison are Veterans Administration hospitals; Forts Robinson and George Wright are state parks; and the others house colleges.

22. RG 94, Lettters received by the commission branch of the Adjutant General's Office, 1863–1870, microcopy 1064, roll 170, NAB; *Army and Navy Journal* 40 (7 March 1903): 663.

23. *ARSW* (1903), 26; Risch, 491; *Annual Report of the Quartermaster General* (1904), 1; and (1906), 56 (hereafter cited as *ARQG*).

24. *ARQGl* (1905), 4; (1907), 22; and (1909), 24; *ARSW* (1910), 280. The army also experimented with the "cement gun" method. *Army and Navy Journal* 49 (27 January 1912): 656.

25. *ARQG* (1905), 7; Lee, *Architects to the Nation,* 194; Wheaton obituary, *New York Times,* 17 March 1931; *ARSW* (1910), 281, 261.

26. Costs listed on building inventories created and consistently updated by each fort, Fort D. A. Russell's at F. E. Warren Air Force Base, and Fort Mackenzie's courtesy of Kevin O'Dell. Lieutenants' double quarters were Quartermaster General's Office (QMGO) plan 120-C, now Buildings Nos. 18 and 20 at Fort D. A. Russell; captains' double quarters were QMGO plan 90, Buildings Nos. 10 and 12. Received from Constructing Quartermaster, 29 February 1904, RG 393, Part V, Entry 27, 427, NAB. Two-company barracks were QMGO plan 75-G; artillery barracks were QMGO plan 150, Buildings Nos. 223 and 224. Received from Constructing Quartermaster, 4 April 1904, RG 393, Part V, Entry 27, 442, NAB; Received from Constructing Quartermaster, 23 November 1904, RG 393, Part V, Entry 27, 50, NAB; *ARQG* (1908), 30; *ARSW* (1908), 39–40.

27. *R. L. Polk's Cheyenne City Directory,* 169; Bartlett, ed., *History of Wyoming,* 2:230, 233. Much of this is repeated verbatim in Keefe's obituary, *Wyoming State Tribune and Cheyenne State Leader,* 14 March 1929.

28. Building No. 21, QMGO plan 129.

29. Buildings Nos. 18 and 20, QMGO plan 120-C. Received from Chief Quartermaster, 14 November 1900, RG 393, Part V, Entry 27, 183–184, NAB; Grashof, "Standardized Plans," 3:120-C.

30. Building No. 12, QMGO plan 90. Grashof, "Standardized Plans," 3:90.

31. Building No. 126, QMGO plan 120-H.

32. Only the double captains' quarters built in 1900 (Building No. 12) and one built two years later (Building No. 10) were built to QMGO plan 90; the rest adhered to QMGO plans 142-A, -B, and -D. They are similar in appearance, with separate porches and recessed entrances. Plan 142-B is entered on the side, in the "back-to-back" arrangement. Plan 142-A has a central cross-gable; this is dropped by plan 142-B, which has two gable dormers, emphasizing the two-unit aspect of the building. The second floor of plan 142-B has four

bedrooms and a bathroom, while the third floor contains two bedrooms and a bathroom for servants. 1908 regulations, p. 177, copy in Clary, "A Life Which Is Gregarious," 670.

33. Building No. 8, QMGO plan 95-A. Keefe received a separate contract for the heating ($1,125), while the electric wiring contract was given to Gilbert, Wilkes & Co. of Denver ($418) and the plumbing contract to Chris Irving for $1,868. Received from Chief Quartermaster, 4 January 1902, RG 393, Part V, Entry 27, 201, NAB; Received from Pope, 3 August 1902, RG 393, Part V, Entry 27, 264, NAB; Received from Constructing Quartermaster, 29 February 1904, RG 393, Part V, Entry 27, 427, NAB.

34. Buildings Nos. 73 and 117, QMGO plan 215.

35. Building No. 92, QMGO plan 3-655 SPL.

36. *Army and Navy Journal* 46 (13 March 1909): 790; and 46 (14 May 1909): 1030.

37. Buildings Nos. 281, 282, and 374–384, QMGO plan 82-D. Building No. 370 was the quarters for civilian employees, QMGO plan 82-P. The standard plans for NCOs were widely used. See Grashof, "Standardized Plans," 1:53–54, 5:NCO-9.

38. Buildings Nos. 223 and 224, QMGO plan 150. Received from Constructing Quartermater, 4 April 1904, RG 393, Part V, Entry 27, 442, NAB; Received from Constructing Quartermaster, 23 November 1904, RG 393, Part V, Entry 27, 50, NAB.

39. Buildings Nos. 220 and 222 for infantry, Buildings Nos. 226 and 228 for cavalry, Buildings Nos. 236 and 238, all to QMGO plans 75-G and 75-M. Received from Constructing Quartermaster, 25 January 1905, RG 393, Part V, Entry 27, 94, NAB.

40. Buildings Nos. 230, 232, 244, 248, 250, QMGO plan 181.

41. The zoning is evident on a site plan produced by the QMGO in January 1917, drawn by C. H. Stone, RG 92, Blueprint File, No. 49, Cartographic Division, National Archives, College Park, Md. This map also indicates the standardized plan numbers for every building, using building numbers still in effect today.

42. Building No. 284, QMGO plan 158. The sale of beer and wine was alternately permitted and prohibited by the War Department. Coffman, *The Old Army,* 360–361, 356.

43. The first four cavalry stables (Buildings Nos. 316, 317, 318 survive) were built in 1906 to QMGO plan 139 and the rest to modifications of this design (plans 139-K and 139-L). Received from Constructing Quartermaster, 19 January 1906, RG 393, Part V, Entry 27, 172, NAB; *ARSW* (1905), 77.

44. Received from Post Adjutant, 30 April 1904, RG 393, Part V, Entry 27, 452, NAB; Received from Chief Quartermaster, 22 May 1904, RG 393, Part V, Entry 27, 456, NAB; Received from Quartermaster General, 10 April 1905, RG 393, Part V, Entry 27. 170, NAB; Received from Post Adjutant, 19 May 1905, RG 393, Part V, Entry 27, 195, NAB; Received from Quartermaster General, 29 March 1906, RG 393, Part V, Entry 27, 206, NAB; Kendall, "History of Fort Francis E. Warren," 44; Adams, *The Post near Cheyenne,* 131; *Annual Report of the Commanding General* (1902), 23.

45. Hansen, "Congressional Career," 6; Adams, *The Post near Cheyenne,* 142, 155; *Army and Navy Journal* 48 (14 January 1911): 560; and 49 (30 September 1911): 120.

46. Contract and specifications with Cheyenne Light, Fuel & Power for purchase of electric current, received from Constructing Quartermaster, 2 July 1904, RG 393, Part V, Entry 29, 6, NAB; "Lighting Plan for Lighting, Wiring, and Installing Fixtures in Buildings at Fort D. A. Russell, Wyo.," dated July 1903, stamped 7 December 1903, RG 92, Blueprint File No. 8, Cartographic Division, National Archives, College, Park, Md. By 1905, Warren owned more than 90 percent of the stock of the Cheyenne Light, Fuel & Power Company. Hansen, "Congressional Career," 6; "History of the Gas, Electric, and Steam Heating Utilities of Cheyenne, Wyoming," no date, Wyoming State Archives, 6–9; *ARQG* (1908), 15; *Annual Report of the Commanding General* (1902), 28.

47. *Army and Navy Journal* 45 (2 November 1907): 215; 45 (38 March 1908): 806; and 47 (11 June 1910): 1220. F. S. Armstrong, "Notice: Instructions for Officers' and N. C. O. Quarters," 29 May 1911, F. E. Warren Air Force Base.

48. Ashburn, *The Elements of Military Hygiene, especially arranged for officers and men of the line,* 20, 23, 56–57, 59, 61–63; *Army and Navy Journal* 47 (18 September 1909): 58; *ARQG* (1905), 9–10; *ARSW* (1904), 294.

49. Coffman, *The Old Army,* 371, 330–11; Millard, "The Shame of Our Army," 412.

50. Millard, "The Shame of Our Army," 414–415, 420.

51. Millard, "The Shame of Our Army," 416; Adams, *The Post near Cheyenne,* 123, 130.

52. "The Point of View," *Scribner's* 52 (1912): 251.

53. Skowronek, *The New American State,* 165; Wiebe, *The Search for Order, 1877–1920,* xiv.

EPILOGUE

1. The name D. A. Russell was then assigned to Camp Marfa, Texas.

2. Adams, *The Post near Cheyenne,* 209; Kendall, "History of Fort Francis E. Warren," 4; Holland, ed., "From Mules to Missiles," 8–1.

3. Adams, *The Post near Cheyenne,* 215, 221.

4. Douglas C. McChristian describes Fort Laramie's post-Army experience in "Fort Laramie: After the Army," "Part I, The Auction," 12–23, and "Part II, The Community," 20–40. For more on the development of Forts Laramie and Bridger as historic sites, see my article "Architecture and Interpretation at Forts Laramie and Bridger," 27–54.

5. The monument bore a 1913 date, but it was not dedicated until 17 June 1915. Fort Bridger received a similar monument in 1914, reading "Established as a trading post in 1834. U.S. Military Post on the Overland Trail June 10, 1858 to October 6, 1890." Jording, *A Few Interested Residents,* 189.

6. Mattes, "Fort Laramie Park History, 1834–1977," 67–68, 80; Greenburg, "Preserving Our Landmarks," 290.

7. Greenburg, "Preserving Our Landmarks," 290; *Cheyenne Daily Leader,* 22 May 1895.

8. *Evanston Herald,* 21 March 1984 (clipping in Wyoming State Archives). Tom Lindmier, superintendent from 1979 to 1987, called it "the worst mistake I ever made," interviewed 3 July 1996, South Pass City, Wyoming; Linda Byers, superintendent, interviewed by phone, 21 September 1998.

9. According to the Historic Sites Act of 1935, public funds could not be used to acquire historic sites for the National Park Service, but if the U.S. government were given the land, the Park Service would preserve the buildings and operate the site. Hosmer, *Preservation Comes of Age,* 1:670–671; Mattes, "Fort Laramie Park History," 104–112, 52, 129; Miller to Rymill, 19 November 1936, Fort Laramie National Historic Site.

10. Proclamation (No. 2292) of 16 July 1938, Act of 29 April 1960 (74 Stat. 83), and P.L. 95-625 (92 Stat. 3487), 10 November 1978, reproduced in "Final Environmental Impact Statement, General Management Plan, Development Concept Plan, Interpretive Prospectus: Fort Laramie National Historic Site," 117–119.

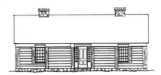

BIBLIOGRAPHY

MANUSCRIPT COLLECTIONS

Washington, D.C.

Library of Congress. Manuscripts Division: Elizabeth J. Burt Papers. Prints and Photographs Division: Historic American Buildings Survey collection. Subject files.

National Archives. Textual Reference Division: RG92, Quartermaster's Department. Entry 225, Consolidated Correspondence File, 1794–1915. Entry 1057, Construction and Repair, 1819–1912. Entry 1213, Correspondence, plans and other papers relating to historical sketches of reservations, 1888–89. Entry 1214, Narrative histories of posts, camps and stations, ca. 1890. RG94, Adjutant General's Office. Letters Received, 1861–1870. Microcopy No. 619, roll 449. Microcopy No. 711 (incl. 1858, 1859), rolls 30, 31. RG112, Surgeon General's Office. Entry 63, Circulars and Circular Letters, 1861–65. RG393, U.S. Army Continental Commands, 1821–1920. Part I, Entry 3898, Department of the Platte, QM Dept. Part V [Records sent from the post], Entry 25, press copies, letters sent, QM Dept. Entry 27, letters received QM Dept. Entry 28, letters sent, QM Dept. Entry 29, letter and telegrams received, QM Dept.

National Park Service. National Register of Historic Places. Files. National Historic Landmarks. Files.

College Park, Maryland

National Archives. Cartographic Division: RG77, Office of the Chief Engineer. RG92, Quartermaster's Department.
Still Pictures Division: RG111, Signal Corps. RG92, Quartermaster's Department.

Cheyenne, Wyoming

F. E. Warren Air Force Base. Files. Inventory, begun in 1905. Architectural drawings.
Wyoming State Archives. Files.
Wyoming State Parks and Historic Sites Division. Fort Bridger files.

Fort Laramie, Wyoming

Fort Laramie National Historic Site. Photographic files. Vertical files. Census data. Historic structures reports. McDermott files. Furnishing plans.

Laramie, Wyoming

University of Wyoming. American Heritage Center. Merrill John Mattes collection. Fort Laramie vertical file. Carter collection.

REPORTS

"Architectural Survey: F. E. Warren Air Force Base." Air Force Regional Civil Engineer, Ballistic Missile Support, 1984.

Beck, Andy. "Historic Structure Report: Architectural Data Section Supplement: 1874 Cavalry Barracks, Fort Laramie National Historic Site, Wyoming." National Park Service, Denver Service Center, 1980.

Brown, William L., III. "A Pictorial History of Enlisted Men's Barracks of the U.S. Army 1861–1895." National Park Service, Harpers Ferry Center, 1984.

Clary, David A. "A Life Which Is Gregarious in the Extreme: A History of Furniture in Barracks, Hospitals, and Guardhouses of the United States Army, 1880–1945." National Park Service, Harpers Ferry Center, 1983.

———. "These Relics of Barbarism: A History of Furniture in Barracks and Guardhouses of the United States Army, 1800–1880." National Park Service, Harpers Ferry Center, no date.

"Final Environmental Impact Statement, General Management Plan, Development Concept Plan, Interpretive Prospectus: Fort Laramie National Historic Site." National Park Service, 1993.

Goodwin, R. Christopher, and Associates. "National Historic Context for Department of Defense Installations, 1790–1940." U.S. Army Corps of Engineers, Baltimore District, 1995.

Grashof, Bethany C. "Standardized Plans, 1866–1940: A Study of United States Army Family Housing." Georgia Institute of Technology, 1986.

Grassick, Mary K. "Historic Furnishings Report: Commanding Officer's Quarters HS-8, Fort Larned National Historic Site." National Park Service, Harpers Ferry Center, 1996.

Holland, Stephen L., ed. "From Mules to Missiles: A History of Francis E. Warren Air Force Base and Its Predecessors." F. E. Warren Air Force Base, 1987.

Mattes, Merrill J. "Fort Laramie Park History, 1834–1977." National Park Service, Rocky Mountain Regional Office, 1980.

McDermott, John D., and James Sheire. "1874 Cavalry Barracks, Fort Laramie National Historic Site, Historic Structures Report, Historical Data Section." National Park Service, Office of History and Historic Architecture, 1970.

Oliva, Leo E. "Fort Union and the Frontier Army in the Southwest." National Park Service, Southwest Cultural Resources Center, 1993.

Pope, Charles, and James W. Sheire. "Historic Structures Report, Part II: The 1883 New Bakery, HB #115, Fort Laramie National Historic Site, Wyoming." National Park Service, Office of Archeology and Historic Preservation, 1969.

Rickey, Don, Jr., and James W. Sheire. "The Cavalry Barracks: Fort Laramie Furnishing Study." National Park Service, Division of History, 1969.

Sheire, James W. "Fort Larned National Historic Site, Historic Structure Report, Part II: Historical Data Section." National Park Service, Office of Archeology and Historic Preservation, 1969.

———. "The 1876 Old Bakery, Fort Laramie National Historic Site, Wyoming, Historic Structure Report, Historical Data Section, Part II." National Park Service, Office of Archeology and Historic Preservation, 1968.

———. "Historic Structure Report, Part I, Non-commissioned Staff Officers' Quarters, Fort Laramie National Historic Site, Wyoming." National Park Service, 1968.
Tetra Tech, Inc. "Fort D. A. Russell/F. E. Warren: Peacekeeper Program Cultural Resources Technical Report 4." U.S. Air Force, 1987.
Wheaton, Rodd. "Lime-Grout and Fort Laramie, Wyoming." National Park Service, 1975(?).

THESES AND DISSERTATIONS

Bowen, Wesley Donald. "The Congressional Career of Senator Francis E. Warren, 1912–1920." M.A. thesis, University of Wyoming, 1949.
Bruegmann, Robert. "Architecture of the Hospital: 1770–1870: Design and Technology." Ph.D. diss., University of Pennsylvania, 1976.
Dubois, William Robert, III. "A Social History of Cheyenne, Wyoming, 1875–1885." M.A. thesis, University of Wyoming, 1963.
Griess, Thomas E. "Dennis Hart Mahan: West Point Professor and Advocate of Military Professionalism, 1830–1871." PhD diss., Duke University, 1968.
Gwilliam, Willard Edward. "A Study of American Army Transitional Construction Process during the Nineteenth Century Trans-Mississippi Frontier Period." M.A. thesis, George Washington University, 1981.
Jones, Robert Franklin. "The Political Career of Senator Francis E. Warren, 1902–1912." M.A. thesis, University of Wyoming, 1949.
Levy, Richard Michael. "The Professionalization of American Architects and Civil Engineers, 1865–1917." Ph.D. diss., University of California, Berkeley, 1980.
O'Dell, Kevin C. "Building Technology in the Department of the Platte, 1866–1890." M.S. thesis, Michigan Technological University, 1997.
Phillips, Davis John. "An Analytical Study for Preserving U.S. Army Barracks Constructed between 1880 and 1940." M.A. thesis, Goucher College, 2000.
Samson, Walter L., Jr. "The Political Career of Senator Francis E. Warren, 1920–1929." M.A. thesis, University of Wyoming, 1951.
Stelter, Gilbert Arthur. "The Urban Frontier: A Western Case Study, Cheyenne, Wyoming, 1867–1887." Ph.D. diss., University of Alberta, 1968.

NEWSPAPERS

Army and Navy Journal (New York)
Army and Navy Register (Washington, D.C.)
Cheyenne Daily Leader (Cheyenne, Wyoming)

BOOKS AND ARTICLES

Adams, Gerald M. *The Post near Cheyenne: A History of Fort D. A. Russell, 1867–1930.* Boulder, Colo.: Pruett Publishing, 1989.
Adams, Judith. *Cheyenne: City of Blue Sky.* Northridge, Calif.: Windsor Publications, 1988.
Allen, John R., and J. H. Walker. *Heating and Ventilation.* New York: McGraw-Hill, 1922.
Alter, J. Cecil. *Jim Bridger.* Norman: University of Oklahoma Press, 1962.
Ambrose, Stephen E. *Duty, Honor, Country: A History of West Point.* Baltimore: Johns Hopkins University Press, 1966.
Anderson, Thomas M. "Army Posts, Barracks and Quarters." *Journal of the Military Service Institution of the United States* 2 (1882): 421–447.
Archer, John. "Country and City in the American Romantic Suburb." *Journal of the Society of Architectural Historians* 42, no. 2 (May 1983): 139–156.
Ashburn, P. M. *The Elements of Military Hygiene, especially arranged for officers and men of the line.* Boston: Houghton Mifflin, 1909.

———. *A History of the Medical Department of the United States Army.* Boston: Houghton Mifflin, 1929.

Athearn, Robert G. *Forts of the Upper Missouri.* Lincoln: University of Nebraska Press, 1967.

———. *High Country Empire: The High Plains and Rockies.* New York: McGraw-Hill, 1960.

———. *Union Pacific Country.* Lincoln: University of Nebraska Press, 1971.

———. *William Tecumseh Sherman and the Settlement of the West.* Norman: University of Oklahoma Press, 1956.

Attebery, Jennifer Eastman. *Building with Logs: Western Log Construction in Context.* Moscow: University of Idaho Press, 1998.

Bandel, Eugene. *Frontier Life in the Army, 1854–1861.* Ed. Ralph P. Bieber. Glendale, Calif.: Arthur H. Clark Co., 1932.

Banham, Reyner. *The Architecture of the Well-Tempered Environment.* Chicago: University of Chicago Press, 1969.

Bartlett, I. S., ed. *History of Wyoming.* Chicago: S. J. Clarke, 1918.

Beard, Frances Birkhead. *Wyoming: From Territorial Days to the Present.* 3 vols. Chicago: American Historical Society, 1933.

Beardslee, Clarence G. "Development of Army Camp Planning." *Civil Engineering* 12, no. 9 (September 1942): 489–492.

Beck, Warren A., and Ynez D. Haase. *Historical Atlas of the American West.* Norman: University of Oklahoma Press, 1989.

Bederman, Gail. *Manliness and Civilization: A Cultural History of Gender and Race in the United States, 1880–1917.* Chicago: University of Chicago Press, 1995.

Beecher, Henry Ward. *Norwood: or, Village Life in New England.* New York: Charles Scribner, 1868.

Bell, George. *Notes on Bread Making, Permanent and Field Ovens, and Bake Houses.* Prepared by Direction of Commissary-General of Subsistence. Washington: GPO, 1882.

Bernardo, C. Joseph, and Eugene H. Bacon. *American Military Policy: Its Development since 1775.* Harrisburg, Pa.: Military Service Publishing, 1955.

Billings, John S. "Hospital Construction and Organization." In *Hospital Plans: Five Essays Relating to the Construction, Organization and Management of Hospitals, contributed by their authors for the use of the Johns Hopkins Hospital of Baltimore.* New York: William Wood and Co., 1875.

———. *The Principles of Ventilation and Heating and Their Practical Application.* New York: The Sanitary Engineer, 1884.

———. *Ventilation and Heating.* New York: The Engineering Record, 1893.

Blackburn, Forrest R. "Army Families in Frontier Forts." *Military Review* (October 1969): 17–28.

Bourke, John G. *On the Border with Crook.* 1891; rpt., Lincoln: University of Nebraska Press, 1971.

Brackett, Albert G. "Fort Bridger: An Origanal [sic] Manuscript Written in 1870." *Wyoming Historical Collections* (1920): 111–120.

———. *History of the United States Cavalry, from the formation of the federal government to the 1st of June, 1863.* New York: Greenwood Press, 1968.

"Brigading the Army." *The Nation* 83 (12 July 1906): 27.

Brown, Dee. *Hear That Lonesome Whistle Blow: Railroads in the West.* New York: Touchstone, 1977.

Brown, Dona. *Inventing New England: Regional Tourism in the Nineteenth Century.* Washington, D.C.: Smithsonian Institution Press, 1995.

Brown, William L., III. *The Army Called It Home: Military Interiors of the Nineteenth Century.* Gettysburg, Pa.: Thomas Publications, 1992.

Bruegmann, Robert. "Central Heating and Forced Ventilation: Origins and Effects on Architectural Design." *Journal of the Society of Architectural Historians* 37 (October 1978): 143–161.

Buecker, Thomas R. "The 1887 Expansion of Fort Robinson." *Nebraska History* 68 (Summer 1987): 83–93.
———. "Fort Niobrara, 1880–1906: Guardian of the Rosebud Sioux." *Nebraska History* 65 (Fall 1984): 301–325.
———. *Fort Robinson and the American West, 1874–1899*. Lincoln: Nebraska State Historical Society, 1999.
———. "The Post of North Platte Station, 1867–1878." *Nebraska History* 63 (Fall 1982): 381–398.
Burlingame, Merrill G. "The Influence of the Military in the Building of Montana." *Pacific Northwest Quarterly* 29 (1938): 135–150.
Bushman, Richard L. *The Refinement of America: Persons, Houses, Cities*. New York: Vintage Books, 1992.
Butler, Ann M. "Military Myopia: Prostitution on the Frontier." *Prologue* 13 (Winter 1981): 233–250.
Calloway, Colin G., ed. *Our Hearts Fell to the Ground: Plains Indians Views of How the West Was Lost*. Boston: St. Martin's Press, 1996.
Carpenter, Rolla C. *Heating and Ventilating Buildings: An Elementary Treatise*. New York: John Wiley and Sons, 1895.
Carr, E. T. "Reminiscences concerning Fort Leavenworth in 1855–56." *Kansas State Historical Society Collections* 12 (1911–1912): 375–383.
Carrington, Frances C. *My Army Life: A Soldier's Wife at Fort Phil Kearny*. 1910; rpt., Boulder, Colo.: Pruett Publishing, 1990.
Carrington, Margaret Irvin. *Absaraka, Home of the Crows: Being the Experience of an Officer's Wife on the Plains*. 1868; rpt., Lincoln: University of Nebraska Press, 1983.
Carter, Thomas. "Traditional Design in an Industrial Age: Vernacular Domestic Architecture in Victorian Utah." *Journal of American Folklore* 104 (Fall 1991): 419–442.
Carter, Thomas, ed. *Images of an American Land: Vernacular Architecture in the Western United States*. Albuquerque: University of New Mexico Press, 1997.
Carter, Thomas, and Bernard L. Herman, eds. *Perspectives in Vernacular Architecture, III*. Columbia: University of Missouri Press, 1989.
Carter, William A. "Fort Bridger in the Seventies." *Annals of Wyoming* 11, no. 2 (April 1939): 111–113.
Centennial Historical Committee. *Cheyenne: The Magic City of the Plains*. Privately printed, 1967.
Chambers, Alex. Letter to H. H. Bancroft, 4 January 1885. *Annals of Wyoming* 5, nos. 2–3 (October 1927–January 1928): 91–95.
Chandler, Alfred D., Jr. *The Visible Hand: The Managerial Revolution in American Business*. Cambridge, Mass.: Harvard University Press, 1977.
Chapman, Carleton B. *Order Out of Chaos: John Shaw Billings and America's Coming of Age*. Boston: Boston Medical Library, 1994.
"Cheyenne." *Collections of the Wyoming Historical Society* 1 (1897): 197–212.
Clary, David A. "The Role of the Army Surgeon in the West: Daniel Weisel at Fort Davis, Texas, 1868–1872." *Western Historical Quarterly* 3, no. 1 (January 1972): 53–66.
Clough, Wilson O., ed. "Fort Russell and Fort Laramie Peace Commission in 1867." Sources of Northwest History, No. 14. Reprinted from *The Frontier* 11, no. 2 (January 1931).
Coffman, Edward M. *The Old Army: A Portrait of the American Army in Peacetime, 1784–1898*. New York: Oxford University Press, 1986.
Comer, Douglas C. *Ritual Ground: Bent's Old Fort, World Formation, and the Annexation of the Southwest*. Berkeley: University of California Press, 1997.
Conforti, Joseph A. *Imagining New England: Explorations of Regional Identity from the Pilgrims to the Mid-Twentieth Century*. Chapel Hill: University of North Carolina Press, 2001.

Connolly, C. P. "Senator Warren of Wyoming." *Colliers Weekly* 49, no. 24 (31 August 1912): 10–11, 30.
Cox, John E. *Five Years in the United States Army: Reminiscences and Record of an Ex-Regular*. Owensville, Md.: General Baptist Publishing House, 1892.
Crellin, J. K. "Airborne Particles and the Germ Theory, 1860–1880." *Annals of Science* 22 (1966): 49–60.
Cronon, William. *Nature's Metropolis: Chicago and the Great West*. New York: W. W. Norton, 1991.
Cronon, William, George Miles, and Jay Gitlin. *Under an Open Sky: Rethinking America's Western Past*. New York: W. W. Norton, 1992.
Cullum, George W. *Biographical Register of the Officers and Graduates of the U. S. Military Academy*. Boston: Houghton Mifflin, 1891.
Culpin, Alan. "A Brief History of Social and Domestic Life among the Military in Wyoming, 1849–1890." *Annals of Wyoming* 45, no. 1 (Spring 1973): 93–108.
Cunliffe, Marcus. *Soldiers and Civilians: The Martial Spirit in America 1775–1865*. New York: Free Press, 1968, 1973.
Custer, Elizabeth B. *"Boots and Saddles": or, Life in Dakota with General Custer*. 1885; rpt., Williamstown, Mass.: Corner House, 1969.
Davenport, B. T. "Horse Cavalry on the Move in 1913." *Periodical: Journal of America's Military Past* 23, no. 4 (Winter 1996): 52–62.
Davis, W. N., Jr. "The Sutler at Fort Bridger." *Western Historical Quarterly* 2, no. 1 (January 1971): 37–54.
Dobak, William A. *Fort Riley and Its Neighbors: Military Money and Economic Growth, 1853–1895*. Norman: University of Oklahoma Press, 1998.
Donaldson, Barry, and Bernard Nagengast. *Heat and Cold: Mastering the Great Indoors*. Atlanta: ASHRAE, 1994.
Downing, Andrew Jackson. *Rural Essays*. New York: Leavitt and Allen, 1857.
Duffy, John. *The Sanitarians: A History of American Public Health*. Urbana: University of Illinois Press, 1990.
Edens, Walter. "Wyoming's Fort Libraries—The March of Intellect." *Annals of Wyoming* 51, no. 2 (Fall 1979): 54–62.
"Efficiency in the Army." *The Nation* 93 (21 December 1911): 595–596.
Ehrenhard, John E. "The Rustic Hotel, Fort Laramie National Historic Site, Wyoming." *Historical Archeology 1973* 7 (1973): 11–29.
Ellison, Robert S. *Fort Bridger, Wyoming: A Brief History*. 1931; rpt., Cheyenne: Wyoming State Archives, Museums and Historical Department, 1981.
Fellman, Michael. *Citizen Sherman: A Life of William Tecumseh Sherman*. New York: Random House, 1995.
Fine, Lenore, and Jesse A. Remington. *The Corps of Engineers: Construction in the United States*. U.S. Army, Office of the Chief of Military History. Washington: GPO, 1972.
Fisch, Arnold G., Jr., and Robert K. Wright, Jr. *The Story of the Noncommissioned Officer Corps: The Backbone of the Army*. Washington: U.S. Army, Center of Military History, 1989.
Fitzgerald, Emily McCorkle. *An Army Doctor's Wife on the Frontier: The Letters of Emily McCorkle Fitzgerald from Alaska and the Far West, 1874–1878*. 1962; rpt., Lincoln: University of Nebraska Press, 1986.
Fogelson, Robert M. *America's Armories: Architecture, Society, and Public Order*. Cambridge, Mass.: Harvard University Press, 1989.
Foner, Jack D. *The United States Soldier between Two Wars: Army Life and Reforms, 1865–1898*. New York: Humanities Press, 1970.
Forman, Sidney. *West Point: A History of the United States Military Academy*. New York: Columbia University Press, 1950.

Forsyth, George A. *The Story of the Soldier*. New York: D. Appleton and Co., 1900.
"Fort Laramie." *Annals of Wyoming* 9, no. 3 (January 1933): 752–754.
Fowler, Orson S. *The Octagon House: A Home for All*. New York: Dover, 1983. Reprint of *A Home for All, or the Gravel Wall and Octagon Mode of Building* (1853).
Foy, Jessica, and Thomas J. Schlereth, eds. *American Home Life, 1880–1930: A Social History of Spaces and Services*. Knoxville: University of Tennessee Press, 1992.
Frazer, Robert W. *Forts of the West: Military Forts and Presidios and Posts Commonly Called Forts West of the Mississippi River to 1898*. Norman: University of Oklahoma Press, 1965.
Frederickson, George M. *The Inner Civil War: Northern Intellectuals and the Crisis of the Nation*. New York: Harper and Row, 1965.
Freeman, Mary E. Wilkins. *A New England Nun and Other Stories*. 1891; rpt. Ridgewood, N. J.: Gregg Press, 1967.
Fry, James B. *Operations of the Army under Buell, from June 10th to October 30th, 1862, and the "Buell Commission."* New York: D. Van Nostrand, 1884.
Garvin, James L. "Mail-Order House Plans and American Victorian Architecture." *Winterthur Portfolio* 16, no. 4 (Winter 1981): 309–334.
Gillett, Mary C. *The Army Medical Department, 1818–1865*. Washington: U.S. Army, Center of Military History, 1987.
Glassie, Henry. *Folk Housing in Middle Virginia: A Structural Analysis of Historic Artifacts*. Knoxville: University of Tennessee Press, 1975.
Gould, Lewis L. "Francis E. Warren and the Johnson County War." *Arizona and the West* 9, no. 2 (Summer 1967): 131–142.
Gowans, Fred R., and Eugene E. Campbell. *Fort Bridger: Island in the Wilderness*. Provo, Utah: Brigham Young University Press, 1975.
Graham, Roy Eugene. "Federal Fort Architecture in Texas during the Nineteenth Century." *Southwestern Historical Quarterly* 74, no. 2 (October 1970): 165–188.
Greenburg, D. W. "Preserving Our Landmarks." *Annals of Wyoming* 6, no. 4 (April 1930): 289–291.
Greene, Francis V. "The United States Army." *Scribner's* 30 (1901): 286–311, 446–462, 593–613.
Greene, Jerome A. "Army Bread and Army Mission on the Frontier, with special reference to Fort Laramie, Wyoming, 1865–1890." *Annals of Wyoming* 47, no. 2 (Fall 1975): 191–219.
Greene, Jerome A., comp. and ed. *Lakota and Cheyenne: Indian Views of the Great Sioux Wars, 1876–77*. Norman: University of Oklahoma Press, 1994.
Grier, Katherine C. *Culture and Comfort: Parlor Making and Middle-Class Identity, 1850–1930*. Washington, D.C.: Smithsonian Institution Press, 1997.
Guentzel, Richard. "The Department of the Platte and Western Settlement, 1866–1877." *Nebraska History* 56, no. 3 (Fall 1975): 389–417.
Haas, Irvin. *Citadels, Ramparts and Stockades: America's Historic Forts*. New York: Everest House, 1979.
Halleck, H. Wager. *Elements of Military Art and Science*. 1846; rpt., New York: D. Appleton and Co., 1860.
Hafen, LeRoy R., and Francis Marion Young. *Fort Laramie and the Pageant of the West, 1834–1890*. Lincoln: University of Nebraska Press, 1938.
Hampton, H. Duane. "The Army and the National Parks." *Montana: The Magazine of Western History* (Summer 1972): 64–79.
Handlin, David P. *The American Home: Architecture and Society, 1815–1915*. Boston: Little, Brown, 1979.
Hansen, Anne Carolyn. "The Congressional Career of Senator Francis E. Warren from 1890 to 1902." *Annals of Wyoming* 20, no. 1 (January 1948): 3–49, and 20, no. 2 (July 1948): 131–158.

Hart, Herbert M. *Old Forts of the Far West.* New York: Bonanza Books, 1965.
———. *Old Forts of the Northwest.* Seattle: Superior Publishing, 1963.
Hedren, Paul L. *Fort Laramie in 1876: Chronicle of a Frontier Post at War.* Lincoln: University of Nebraska Press, 1988.
Heitman, Francis B. *Historical Register and Dictionary of the United States Army.* Washington, D.C.: GPO, 1903.
Hoagland, Alison K. "Architecture and Interpretation at Forts Laramie and Bridger." *The Public Historian* 23, no. 1 (Winter 2001): 27–54.
———. "'The Invariable Model': Standardization and Military Architecture in Wyoming, 1860–1890." *Journal of the Society of Architectural Historians* 57, no. 3 (September 1998): 298–315.
———. "Village Constructions: U.S. Army Forts on the Plains, 1848–1890." *Winterthur Portfolio* 34, no. 4 (Winter 1999): 215–237.
Hogg, Ian V. *Fortress: A History of Military Defence.* New York: St. Martin's Press, 1975.
Holtz, Milton E. "Old Fort Kearny, 1846–1848: Symbol of a Changing Frontier." *Montana: The Magazine of Western History* (Autumn 1972): 44–55.
Hosmer, Charles B. *Preservation Comes of Age: From Williamsburg to the National Trust, 1926–1949.* 2 vols. Charlottesville: University of Virginia Press, 1981.
Hospital Plans: Five Essays Relating to the Construction, Organization and Management of Hospitals, contributed by their authors for the use of the Johns Hopkins Hospital of Baltimore. New York: William Wood & Co., 1875.
Hume, Edgar Erskine. "John Shaw Billings as an Army Medical Officer." *Bulletin of the History of Medicine* 6 (1938): 225–270.
Huntington, Samuel P. *The Soldier and the State: The Theory and Politics of Civil-Military Relations.* Cambridge, Mass.: Harvard University Press, 1957.
Hurt, R. Douglas. "The Construction and Development of Fort Wallace, Kansas, 1865–1882." *Kansas Historical Quarterly* 43, no. 1 (1977): 44–55.
Jackson, John Brinckerhoff. *American Space: The Centennial Years, 1865–1876.* New York: W. W. Norton, 1972.
Jamieson, Perry D. *Crossing the Deadly Ground: United States Army Tactics, 1865–1899.* Tuscaloosa: University of Alabama Press, 1994.
Jeffrey, Julie Roy. *Frontier Women: The Trans-Mississippi West, 1840–1880.* New York: Hill and Wang, 1979.
Jennings, Jan. "Frank J. Grodavent: Western Army Architect." *Essays in Colorado History* 11 (1990): 1–23.
Jensen, Robert. "Board and Batten Siding and the Balloon Frame: Their Incompatibility in the Nineteenth Century." *Journal of the Society of Architectural Historians* 30, no. 1 (March 1971): 40–50.
Jewett, Sarah Orne. *Country By-Ways.* 1881; rpt., Freeport, N. Y.: Books for Libraries Press, 1969.
Johnson, Hervey. *Tending the Talking Wire: A Buck Soldier's View of Indian Country, 1863–1866.* Ed. William E. Unrau. Salt Lake City: University of Utah Press, 1979.
Johnson, Susan Lee. "'A Memory Sweet to soldiers': The Significance of Gender in the History of the 'American West.'" *Western Historical Quarterly* 24, no. 4 (November 1993): 495–517.
Jording, Mike. *A Few Interested Residents: Wyoming Historical Markers and Monuments.* Privately printed, 1992.
Junge, Mark. *J. E. Stimson: Photographer of the West.* Lincoln: University of Nebraska Press, 1985.
Karsten, Peter, ed. *The Military in America: From the Colonial Era to the Present.* New York: Free Press, 1986.
Kemble, C. Robert. *The Image of the Army Officer in America: Background for Current Views.* Westport, Conn.: Greenwood Press, 1973.

Kendall, Jane R. "History of Fort Francis E. Warren." *Annals of Wyoming* 18, no. 1 (January 1946): 3–66.
King, Charles. *Captain Blake*. Philadelphia: J. B. Lippincott, 1891.
———. *"Laramie"; or, the Queen of Bedlam*. 1889; rpt., Fort Laramie Historical Association, 1986.
Kirkus, Peggy Dickey. "Fort David A. Russell: A Study of Its History from 1867 to 1890." *Annals of Wyoming* 40, no. 2 (October 1968): 161–192, and 41, no. 1 (April 1969): 83–111.
Klauss, A. K., R. H. Tull, L. M. Roots, and J. R. Pfafflin. "History of the Changing Concepts in Ventilation Requirements." *ASHRAE Journal* 12 (June 1970): 51–55.
Knight, Oliver. *Life and Manners in the Frontier Army*. Norman: University of Oklahoma Press, 1978.
Lamers, William M. *Edge of Glory: A Biography of General William S. Rosecrans, U.S.A.* New York: Harcourt, Brace and World, 1961.
Lankton, Larry. *Cradle to Grave: Life, Work, and Death at the Lake Superior Copper Mines*. New York: Oxford University Press, 1991.
Laramie County Historical Society. *Early Cheyenne Homes, 1880–1890*. 2nd ed. Privately printed, 1964.
Larson, Magali Sarfatti. *The Rise of Professionalism: A Sociological Analysis*. Berkeley: University of California Press, 1977.
Larson, T. A. *History of Wyoming*. 2nd ed. Lincoln: University of Nebraska Press, 1978.
Lavender, David. *Fort Laramie and the Changing Frontier*. NPS Handbook 118. Washington, D.C.: GPO, 1983.
Leckie, Shirley. "Reading between the Lines: Another Look at Officers' Wives in the Post–Civil War Frontier Army." *Military History of the Southwest* 19, no. 2 (Fall 1989): 137–160.
Lee, Antoinette J. *Architects to the Nation: The Rise and Decline of the Supervising Architect's Office*. New York: Oxford University Press, 2000.
Leopold, Richard W. *Elihu Root and the Conservative Tradition*. Boston: Little, Brown, 1954.
Lewis, Kenneth E. *The American Frontier: An Archaeological Study of Settlement Pattern and Process*. Orlando: Academic Press, 1984.
Limerick, Patricia Nelson. *The Legacy of Conquest: The Unbroken Past of the American West*. New York: W. W. Norton, 1987.
Linn, Brian McAllister. "The Long Twilight of the Frontier Army." *Western Historical Quarterly* 27, no. 2 (Summer 1996): 141–167.
Lombard, Jess H. "Old Bedlam." *Annals of Wyoming* 13, no. 2 (April 1941): 87–91.
Mahan, D. H. *Descriptive Geometry, as applied to the drawing of Fortification and Sterotomy for the use of the Cadets of the U.S. Military Academy*. New York: John Wiley and Son, 1873.
Mansfield on the Condition of the Western Forts, 1853–54. Ed. Robert W. Frazer. Norman: University of Oklahoma Press, 1963.
Mantor, Lyle E. "Fort Kearny and the Westward Movement." *Nebraska History* 29 (September 1948): 175–207.
Marcus, Robert D., and David Burner, eds. *America Firsthand, Volume I: From Settlement to Reconstruction*. 2nd ed. New York: St. Martin's Press, 1992.
Marcy Randolph B. *The Prairie Traveler: A Hand-Book for Overland Expeditions*. New York: Harper & Bros., 1859.
Marshall, Mortimer M., Jr. "From Barracks to Dormitories." *The Military Engineer* 66 (November–December 1974): 343–346.
Massey, Rheba. "The Preservation of Wyoming's Vernacular Architecture." *Annals of Wyoming* 63, no. 4 (Fall 1991): 172–174.
Mattes, Merrill J. *Fort Laramie and the Forty-Niners*. Estes Park, Colorado: Rocky Mountain Nature Association, 1949.

———. *Indians, Infants and Infantry: Andrew and Elizabeth Burt on the Frontier.* 1960; rpt., Lincoln: University of Nebraska Press, 1988.

Matthews, Franklin. "The American Soldier: An Improved Fighter." *Harper's Weekly* 47 (30 May 1903): 894.

Mattison, Ray H. "The Army Post on the Northern Plains, 1865–1895." *Nebraska History* 35 (1954): 1–27.

McChristian, Douglas C. "The Bug Juice War." *Annals of Wyoming* 49, no. 2 (Fall 1977): 253–261.

———. "Fort Laramie—After the Army: Part I, The Auction." *Annals of Wyoming* 73, no. 3 (Summer 2001): 12–23.

———. "Fort Laramie—After the Army: Part II, The Community." *Annals of Wyoming* 73, no. 4 (Autumn 2001): 20–40.

———. "Fort Laramie—After the Army: Part III, Preservation." *Annals of Wyoming* 74, no. 4 (Autumn 2002): 14–31.

———. *The U.S. Army in the West, 1870–1880: Uniforms, Weapons, and Equipment.* Norman: University of Oklahoma Press, 1995.

McDermott, John D. *Fort Mackenzie: A Century of Service.* Sheridan: Fort Mackenzie Centennial Committee, 1998.

———. "Fur Trade and Military History: Wyoming Historiography and the Nineteenth Century." *Annals of Wyoming* 63, no. 4 (Fall 1991): 131–135.

McPhee, John. *Rising from the Plains.* New York: Noonday Press, 1986.

Meinig, D. W. *The Shaping of America: A Geographical Perspective on 500 Years of History.* Vol. 2: *Continental America, 1800–1867.* New Haven: Yale University Press, 1993. Vol. 3: *Transcontinental America, 1850–1915.* New Haven: Yale University Press, 1998.

Meinig, D. W., ed. *The Interpretation of Ordinary Landscapes: Geographical Essays.* New York: Oxford University Press, 1979.

Millard, Bailey. "More about 'The Shame of Our Army.'" *Cosmopolitan Magazine* 49 (November 1910): 758–760.

———. "The Shame of Our Army: Why Fifty Thousand Enlisted American Soldiers Have Deserted." *Cosmopolitan Magazine* 49 (September 1910): 411–420.

———. "The Story of a Deserter." *Cosmopolitan Magazine* 50 (1910): 274–282.

Miller, Darlis A. *Soldiers and Settlers: Military Supply in the Southwest, 1861–1885.* Albuquerque: University of New Mexico Press, 1989.

Miller, Don C., and Stan B. Cohen. *Military and Trading Posts of Montana.* Missoula: Pictorial Histories Publishing Company, 1970.

Millis, Walter. *Arms and Men: A Study in American Military History.* 1956; rpt., New Brunswick, N. J.: Rutgers University Press, 1981.

Milner, Clyde A., II, Carol A. O'Connor, and Martha Sandweiss. *The Oxford History of the American West.* New York: Oxford University Press, 1994.

Mullin, Cora Phoebe. "The Founding of Fort Hartsuff." *Nebraska History Magazine* 12 (April–June 1931): 129–140.

Mumford, Lewis. *Technics and Civilization.* 1934; rpt., New York: Harcourt, Brace, 1963.

Murray, Robert A. *Military Posts in the Powder River Country of Wyoming, 1865–1894.* Lincoln: University of Nebraska Press, 1968.

———. *Military Posts of Wyoming.* Fort Collins, Colo.: Old Army Press, 1974.

———. "Prices and Wages at Fort Laramie, 1881–1885." *Annals of Wyoming* 36, no. 1 (April 1964): 19–21.

Murray, Robert A., ed. *Fort Laramie: "Visions of a Grand Old Post."* Fort Collins, Colo.: Old Army Press, 1974.

Myres, Sandra L. "Frontier Historians, Women, and the 'New' Military History." *Military History of the Southwest* 19, no. 1 (Spring 1989): 27–37.

———. "Romance and Reality on the American Frontier: Views of Army Wives." *Western Historical Quarterly* 13 (October 1982): 409–427.

Nenninger, Timothy K. *The Leavenworth Schools and the Old Army: Education, Professionalism, and the Officer Corps of the United States Army, 1881–1918*. Westport, Conn.: Greenwood Press, 1978.

Noble, Bruce J., Jr. "Marking Wyoming's Oregon Trail." *Overland Journal* (Summer 1986): 19–31.

Notes on Building in Concrete and Pisé for the Frontier. Washington, D.C.: War Department, Quartermaster General's Office, 1873.

Nowak, Timothy R. "From Fort Pierre to Fort Randall: The Army's First Use of Portable Cottages." *South Dakota History* 32, no. 2 (Summer 2002): 95–116.

Ogle, Maureen. *All the Modern Conveniences: American Household Plumbing, 1840–1890*. Baltimore: Johns Hopkins University Press, 1996.

"Old Fort Bridger." *Annals of Wyoming* 11, no. 3 (July 1939): 219–221.

Ostrander, Alson B. *An Army Boy of the Sixties*. Yonkers-on-Hudson, New York: World Book Co., 1926.

Outline Descriptions of the Posts in the Military Division of the Missouri, commanded by Lieutenant General P. H. Sheridan. 1876; rpt., Bellevue, Nebr.: Old Army Press, 1969.

Parkman, Francis. *The Oregon Trail*. 1872; rpt., Boston: Little, Brown, 1910.

Peterson, Charles E., ed. *Building Early America: Contributions toward the History of a Great Industry*. Mendham, N.J.: Astragal Press, 1976.

Pitcaithley, Dwight T. "The Third Fort Union: Architecture, Adobe, and the Army." *New Mexico Historical Review* 57, no. 2 (1982): 123–137.

Pocius, Gerald L., ed. *Living in a Material World: Canadian and American Approaches to Material Culture*. St. John's, Newfoundland: Institute of Social and Economic Research, 1991.

"The Point of View." *Scribner's* 52 (1912): 250–251.

"Post Sutlers at Fort Laramie Nebraska Territory." *Annals of Wyoming* 4, no. 2 (October 1926): 318.

Prucha, Francis Paul. *Broadax and Bayonet: The Role of the U.S. Army in the Development of the U.S. 1815–1860*. Madison: State Historical Society of Wisconsin, 1953.

———. *A Guide to the Military Posts of the United States, 1789–1895*. Madison: State Historical Society of Wisconsin, 1964.

Reed, Henry Hope. *The New York Public Library: Its Architecture and Decoration*. New York: W. W. Norton, 1986.

Rees, Ronald. *New and Naked Land: Making the Prairies Home*. Saskatoon: Western Producer Prairie Books, 1988.

Reid, Richard J. *The Army That Buell Built*. Privately printed, 1994.

Reps, John W. *Cities of the American West: A History of Frontier Urban Planning*. Princeton: Princeton University Press, 1979.

———. *The Making of Urban America: A History of City Planning in the United States*. Princeton: Princeton University Press, 1965.

Richmond, Phyllis Allen. "Some Variant Theories in Opposition to the Germ Theory of Disease." *Journal of the History of Medicine* 9 (1954): 290–303.

Rickey, Don, Jr. *Forty Miles a Day on Beans and Hay: The Enlisted Soldier Fighting the Indian Wars*. Norman: University of Oklahoma Press, 1963.

Risch, Erna. *Quartermaster Support of the Army, 1775–1939*. Washington, D.C.: U.S. Army, Center of Military History, 1962.

Roberts, Phil, ed. *Readings in Wyoming History: Issues in the History of the Equality State*. Laramie: Skyline West, 1993.

Robinson, Willard B. *American Forts: Architectural Form and Function*. Urbana: University of Illinois Press, 1977.

Robrock, David P. "A History of Fort Fetterman, Wyoming, 1867–1882." *Annals of Wyoming* 48, no. 1 (Spring 1976): 5–76.

Roe, Frances M. A. *Army Letters from an Officer's Wife*. 1909; rpt., New York: Arno Press, 1979.

Rusling, James F. *Across America: or, The Great West and the Pacific Coast*. New York: Sheldon and Co., 1875.

Ryan, John P. *Fort Stanton and Its Community, 1855–1896*. Las Cruces, N.M.: Yucca Tree Press, 1998.

Saltiel, E. H., and Geo. Barnett. *History and Business Directory of Cheyenne*. Cheyenne: L. B. Joseph, 1868.

Schlereth, Thomas J., ed. *Material Culture: A Research Guide*. Lawrence: University Press of Kansas, 1985.

Schubert, Frank N. *Outpost of the Sioux Wars: A History of Fort Robinson*. Lincoln: University of Nebraska Press, 1993.

———. *Vanguard of Expansion: Army Engineers in the Trans-Mississippi West, 1819–1879*. Washington, D.C.: GPO, 1980.

Shields, Alice Mathews. "Army Life on the Wyoming Frontier." *Annals of Wyoming* 13, no. 4 (October 1941): 331–343.

Shryock, Richard Harrison. *The Development of Modern Medicine: An Interpretation of the Social and Scientific Factors Involved*. Philadelphia: University of Pennsylvania Press, 1936.

Simonin, Louis L. *The Rocky Mountain West in 1867*. Lincoln: University of Nebraska Press, 1966.

Skowronek, Stephen. *Building a New American State: The Expansion of National Administrative Capacities, 1877–1920*. Cambridge: Cambridge University Press, 1982.

Slotkin, Richard. *Fatal Environment: The Myth of the Frontier in the Age of Industrialization, 1800–1890*. New York: Harper Perennial, 1985.

Smith, Merritt Roe, ed. *Military Enterprise and Technological Change*. Cambridge, Mass.: MIT Press, 1985.

Smith, Sherry L. "Lost Soldiers: Re-searching the Army in the American West." *Western Historical Quarterly* 29, no. 2 (Summer 1998): 149–163.

———. "New Directions for Wyoming Women's History." *Annals of Wyoming* 63, no. 4 (Fall 1991): 150–152.

———. *The View from Officers' Row: Army Perceptions of Western Indians*. Tucson: University of Arizona Press, 1990.

Smith, Thomas T. *The U.S. Army and the Texas Frontier Economy, 1845–1900*. College Station: Texas A&M Press, 1999.

"Soldiering on the Frontier." *Annals of Wyoming* 35, no. 1 (April 1963): 83–84.

Spring, Agnes Wright. *Caspar Collins: The Life and Exploits of an Indian Fighter of the Sixties*. New York: Columbia University Press, 1927.

———. "Old Letter Book." *Annals of Wyoming* 13, no. 4 (October 1941): 237–253.

Spring, Agnes Wright, ed. "An Army Wife Comes West: Letters of Catharine Wever Collins (1863–1864)." *Colorado Magazine* 31, no. 4 (October 1954): 1–33.

Stallard, Patricia Y. *Glittering Misery: Dependents of the Indian Fighting Army*. Norman: University of Oklahoma Press, 1978.

Starr, Eileen F. *Architecture in the Cowboy State, 1849–1940: A Guide*. Glendo, Wyo.: High Plains Press, 1992.

Starr, Paul. *The Social Transformation of American Medicine*. New York: Basic Books, 1982.

Stelter, Gilbert. "The City and Westward Expansion: A Western Case Study." *Western Historical Quarterly* 4, no. 2 (April 1973): 187–202.

Stewart, Miller J. "Army Laundresses: Ladies of the `Soap Suds Row.'" *Nebraska History* 61 (1980): 421–436.

Stowe, Harriet Beecher. *Oldtown Folks*. 1869; rpt., Cambridge, Mass.: Harvard University Press, 1966.

Strahorn, Carrie Adell. *Fifteen Thousand Miles by Stage*. New York: G. P. Putnam's Sons, 1911.

Summerhayes, Martha. *Vanished Arizona: Recollections of the Army Life of a New England Woman*. 1911; rpt., Lincoln: University of Nebraska Press, 1979.
Tate, James. P., ed. *The American Military on the Frontier*. Washington, D.C.: Office of Air Force History, 1978.
Tate, Michael L. *The Frontier Army in the Settlement of the West*. Norman: University of Oklahoma, 1999.
Taylor, George Rogers, ed. *The Turner Thesis: Concerning the Role of the Frontier in American History*. Lexington, Mass.: D. C. Heath and Company, 1956.
Thompson, John D., and Grace Goldin. *The Hospital: A Social and Architectural History*. New Haven: Yale University Press, 1975.
Toole, K. Ross, et al. *Probing the American West: Papers from the Santa Fe Conference*. Santa Fe: Museum of New Mexico Press, 1962.
Triggs, J. H. *History of Cheyenne and Northern Wyoming embracing the Gold Fields of the Black Hills*. Omaha: Herald Steam Book and Job Printing House, 1876.
Truettner, William H., ed. *The West as America: Reinterpreting Images of the Frontier, 1820–1920*. Washington, D.C.: Smithsonian Institution Press, 1991.
U.S. Army Corps of Engineers. *The History of the U.S. Army Corps of Engineers*. Privately printed, 1986.
U.S. War Department. *Regulations concerning Barracks and Quarters for the Army of the United States, 1860*. Washington, D.C.: George W. Bowman, 1861.
———. *Regulations for the Army of the United States, 1861*. New York: Harper and Bros., 1861.
———. *General Regulations for the Army*. Washington, D.C.: GPO, 1876.
———. *Regulations for the Army of the United States*. Washington, D.C.: GPO, 1889.
———. *Regulations for the Army of the United States, 1904*. Washington, D.C.: GPO, 1905.
———. *Regulations for the Army of the United States 1913*. Washington, D.C.: GPO, 1917.
U.S. War Department, Quartermaster-General's Office. *Outline Description of Military Posts and Reservations in the United States and Alaska and of National Cemeteries*. Washington, D.C.: GPO, 1904.
———. *Notes on Building in Concrete and Pisé for the Frontier*. Washington, D.C., 1873.
U.S. War Department, Surgeon-General's Office, Circular No. 4. *A Report on Barracks and Hospitals, with Descriptions of Military Posts*. Washington, D.C.: GPO, 1870.
———. Circular No. 8. *A Report on the Hygiene of the United States Army with Descriptions of Military Posts*. Washington: GPO, 1875.
Unruh, John D., Jr. *The Plains Across: The Overland Emigrants and the Trans-Mississippi West, 1840–60*. Urbana: University of Illinois Press, 1982.
Upton, Dell. *Architecture in the United States*. Oxford: Oxford University Press, 1998.
———. "Pattern Books and Professionalism: Aspects of the Transformation of Domestic Architecture in America." *Winterthur Portfolio* 19, nos. 2–3 (Summer–Autumn 1984): 107–150.
Upton, Dell, ed. *America's Architectural Roots: Ethnic Groups That Built America*. Washington, D.C.: Preservation Press, 1986.
Upton, Dell, and John Michael Vlach, eds. *Common Places: Readings in American Vernacular Architecture*. Athens: University of Georgia Press, 1986.
Utley, Robert. *Frontier Regulars: The United States Army and the Indian, 1866–1891*. Lincoln: University of Nebraska Press, 1973.
———. *Frontiersmen in Blue: The United States Army and the Indian, 1848–1865*. New York: Macmillan, 1967.
———. *Indian, Soldier, Settler: Experiences in the Struggle for the American West*. St. Louis: Jefferson National Expansion Historical Association, 1979.
Vaughn, J. W. "The Fort Laramie Hog Ranches." *The Westerners* 13, no. 2 (1966): 39–41.
Wade, Arthur P. "The Military Command Structure: The Great Plains, 1853–1891." *Journal of the West* 15, no. 3 (July 1976): 5–22.

Ware, Eugene F. *The Indian War of 1864*. 1911; rpt., New York: St. Martin's Press, 1960.
Warner, Ezra J. *Generals in Blue: Lives of the Union Commanders*. Baton Rouge: Louisiana State University Press, 1992.
Weigley, Russell F. *The American Way of War: A History of the United States Military Strategy and Policy*. Bloomington: Indiana University Press, 1973.
———. *History of the United States Army*. New York: Macmillan, 1967.
———. *Quartermaster General of the Union Army: A Biography of M. C. Meigs*. New York: Columbia University Press, 1959.
Wells, Camille, ed. *Perspectives in Vernacular Architecture, II*. Columbia: University of Missouri Press, 1986.
Welty, Raymond L. "The Army Fort of the Frontier (1860–1870)." *North Dakota Historical Quarterly* 2, no. 3 (April 1928): 155–167.
Wheaton, Francis B. "The Architecture of the Army Post." *Quartermaster Review* 8 (September–October 1928): 10–13.
White, Richard. *"It's Your Misfortune and None of My Own": A New History of the American West*. Norman: University of Oklahoma Press, 1991.
Whitman, S. E. *The Troopers: An Informal History of the Plains Cavalry, 1865–1890*. New York: Hastings House, 1962.
Wiebe, Robert H. *The Search for Order, 1877–1920*. New York: Hill and Wang, 1967.
Williams, R. "Army Organization in the United States." *The Galaxy* 24 (November 1877): 594–602.
———. "The Staff of the United States Army." *Atlantic Monthly* 41 (March 1878): 376–384.
Wilson, Chris. *The Myth of Santa Fe: Creating a Modern Regional Tradition*. Albuquerque: University of New Mexico Press, 1997.
Wood, Joseph S. "'Build, Therefore, Your Own World': The New England Village as Settlement Ideal." *Annals of the Association of American Geographers* 81, no. 1 (1991): 32–50.
———. *The New England Village*. Baltimore: Johns Hopkins University Press, 1997.
Wooster, Robert. *The Military and United States Indian Policy, 1865–1903*. New Haven: Yale University Press, 1988.
———. *Nelson A. Miles and the Twilight of the Frontier Army*. Lincoln: University of Nebraska Press, 1993.
Wright, Gwendolyn. *Moralism and the Model Home: Domestic Architecture and Cultural Conflict in Chicago, 1873–1913*. Chicago: University of Chicago Press, 1980.

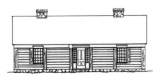

INDEX

References to illustrations are in italic type.

Académie des Sciences, 162
Alden, C. H., 94, 142, 162
Alger, Russell, 206
American Mountain Men Association, 248
American Public Health Association, 158
Anderson, Thomas M., 24, 36, 69, 98, 134–35, 150–51, 185, 190
Architectural styles: classical, 39–40, 211; (English) colonial revival, 211, 221, 223; Georgian, 39, 75, 126; Gothic Revival, 52; Spanish colonial revival, 211
Architecture: construction practices, 58–64, 74–75, 139, 167–71, 212–16; heating, 185–86; lighting, 196–99, 240; prefabrication, 141–42; professionalization, 242; space allocation, 64–66, 134–37; ventilation, 55–56, 160–63, *164*, 180, 183–85. *See also* Building types, Model plans, Standard plans, individual forts
Army Medical Library, 158
Ashburn, P. M., 240
Atlas Intercontinental Ballistic Missile Headquarters, 245
Augur, Christopher C., 86

Bannock Indians, 208
Barber, Amos, 209
Beecher, Henry Ward, 92
Belknap, William W., 18, 74
Bell, George, 153
Billings, John Shaw, 41, 157–66, 180, 183–85, 195, 240
Black & Clark, *212*
Blickensderfer, Jacob, Jr., 86
Bourke, John G., 89, 181
Bozeman Trail, 8, 20, 29
Brackett, Albert G., 195
Bradley, Luther P., 119
Brechemin, Louis, *105*, 187
Bridge across North Platte River, 91, 142–43, *144*
Bridger, Jim, 27, 248
Brown, W. H., 116
Brush Electric Lighting Company, 198
Buchanan, James, 28
Buell, Don Carlos, 41, 157
Buffalo Springs, Tex., 19
Building materials, 66–74, 139–47
Building types: barracks, 50, 179–82, 183–85; bungalow, 227; canteen, 116; chapel, 99; hospital, 153–67; laundresses' quarters, 66; library, 99; noncommissioned officers' quarters, 137; officers' quarters, 65–66, 98, 108, 139, 174; post trader's store, 114–17; shop, 113; tent, 141. *See also* Model plans, Standard plans, individual forts

Bullock, William G., 52, 115, 116
Burt, Andrew S., 119
Burt, Elizabeth, 90, 101, 112, 113, 117, 118, 119–20, 131, 141–42, 177, 178, 187, 196
Bushman, Richard, 8, 40

Caldwell, D. G., 187
Camp Lancaster, Tex., 141
Camp Marfa, Tex., 268n.1
Camp Medicine Butte, Wyo., 88
Camp Pilot Butte, Wyo., 88
Canby, Edward R. S., 71, 179
Capehart Housing Project, 245
Capron, Cynthia, 98, 113
Carey, Joseph M., 193
Carrington, Frances C., 111, 177
Carrington, Henry B., 20, 23, 29, 30, 62, 68, 71–72
Carrington, Margaret, 8, 69, 101, 111, 117
Carter, Mary, 115, 132
Carter, William A., 28, *109*, 114–15, 131–32
Central Pacific Railroad, 82
Cheyenne, Wyo.: brickyards, 140; and Fort D. A. Russell, 86–92, *90*, *94*, 172–73; electricity, 198; and F. E. Warren, 204–205; and M. P. Keefe, 216; water system, 188, 190, 236–37
Cheyenne Electric Railway, 237–38
Cheyenne Indians, 18
Cheyenne Quartermaster Depot, 60, 86, 91, 213
Chittenden, Hiram M., 191–92
Chivington, John M., 18, 33
Chouteau, Pierre, 25
Church of Latter Day Saints. *See* Mormons
Chynoweth, Edward, 66, 169, 193
Civil War, 18, 41, 43
Clement (construction foreman), 168–69
Codgess Construction, 219
Collins, Catharine, 52
Collins, Gilbert H., *55*
Collins, John S., 116
Collins, William O., 45–46
Connolly, C. P., 204–205
Connor, Patrick, 27
Converse, Amasa R., 204
Cook, P. S., 219
Cooke, Philip St. George, 18, 19, 34, 117
Cooperative Building Plan Association, 169
Coxey's Army, 209

Crazy Horse, 83
Crook, George, 83, 174
Crow Indians, 27, 117
Cuny, Adolph, 145
Custer, Elizabeth, 88, 111, 121, 150
Custer, George Armstrong, 83, 85, 113

Dandy, George, 48–49, 62, 64, 65, 150, 168, 190
Daughters of the American Revolution, 247
Davis, Jefferson, 21, 22
Davis, Richard Harding, 106
Demaray, Arthur E., 249
De Trobriand, Philip Regis, 33
Dodge, Grenville, 86
Donaldson, James L., 44, 141, 142
Donelson, Andrew J., 32
Douglass, Henry, 74
Downing, Andrew Jackson, 80, 101
Drew, George A., 60
DuBois, William, 266n.19
Duncan, Thomas, 119
Durfee, Lucius L., 66
Dwyer Plumbing and Heating, 219

Easton, Langdon C., 44
Elliott, B. (contractor), 146
Eltonhead, Francis, 168–69, 179, 190
Escoffey, Jules, 145

Fetterman, William J., 20
FitzGerald, Emily, 175
Flint, Franklin F., 47
Floyd, John B., 21
Forsyth, George, 103
Fort Abraham Lincoln, N. Dak., 88, 113, 121, 150, 153
Fort Belknap, Tex., 19
Fort Boise, Idaho, 175
Fort Bridger, *28*, *35*, *107*, *109*, *193*; bandstand, 102; barracks, 54–56, 56–57, *59*, *60*, 65, 71, 131, 132, *134*, 168–69, 179, 185; blacksmith shop, 56; civilian employees, 113; closing, 132, 203, 245; corral, 35; fences, 104, 106; fortifications, 35; guard house, 132, 168; hospital, 56, 147, 155–57, 164; latrine, 186; laundresses' quarters, 56, 132; layout, 37–38; library, 99; lighting, 196; location of, *15*, 27–29; officers' quarters, 53–54, *56*, 56, *57*, *58*, *71*, 71, 75, 113, 131–32, 150–51, *156*, *157*, 177, 179, 248; oil house, 132, 168; panel construction, 69–72, *70*, *71*;

parade ground, 37–38, 101; post trader, 55, 114–15; post trader's house, 115, 131–32, *133*; post-Army, 248–49; purpose, 9, 27–28; sawmill, 55; soldiers' disease rate, 154–55; storehouse, 53, 56; supply system, 67, 142; teamsters' quarters, 132; as temporary post, 131; trading post, 9, 27, 248–49; warehouses, 132; water distribution, 187, 190–92

Fort C. F. Smith, Mont., 29, 117

Fort Chadbourne, Tex., 19

Fort Crook, Neb., 213, 216

Fort Dallas, Ore., 256n.1

Fort D. A. Russell, *87*, *96*, *97*, *191*, *210*; administration building, 211, *212*, 234; barracks, 74, 95, 119, 122–23, 124, *125*, 125, *126*, *182*, 184, 210, 211, 215, *233*, 233–34, *234*, 244, 245; baseball diamond, 106–107; bath house, 195–96, *197*; bowling alley, 106; Brown's Mess House, 95; chapel, 99; and Cheyenne, 89–92, 172–74; civilian employees, 64, 91, 113; electricity, 198, 238, 240; establishment of, 86–88; fire, 121; guard house, 95, 123, 152, *159*, *160*, 219, 234; guard tower, 123, *124*; gymnasium, 245; hospital, 62, 95, 99, 162–63, 166, *167*, *168*, 210, 245; landscaping, 236–38; latrine, 186; laundresses, 109; laundresses' quarters, 95, 123–24; layout, 92–97, 209–11; location, *15*, 86; mess hall, 144; noncommissioned officers' quarters, 124, 137, *138*, 194, 210, 227, 230–33, *232*, 245; officers' quarters, 95, 97, *102*, 119–21, *120*, 122, *123*, 124, 125–31, *128*, *129*, 135, *136*, 141, 169, 173, 176, 190, 194, 210, 211, 215, 216–27, *217*, *220*, *221*, *223*, *225*, *226*, *227*, *230*, *231*, 244; parade ground, 95, 101, 131, 210; as permanent post, 124; post exchange and gymnasium, 234–35, *238*; post trader's house, 95; post trader's store, 95; purpose, 9, 86, 203–204, 205, 207, 208–209; sanitation, 238; sawmill, 95, 187, 196; school and library, 95; sewer system, 193–95; shops, 95, 123; social event, 172; soldiers, 108; soldiers, African American, 241–42; stables, 95, 123, 210, 235–36; steam laundry, 196; storehouses, 125, 142; supply system, 140; tennis courts, 107; theater, 244; warehouses, 95, 210; water distribution, 187–90, 236–37

Fort Douglas, Utah, 27, 194

Fort Duchesne, Utah, 192

Fort Ethan Allen, Vt., 219

Fort Fetterman, Wyo., 75, 91, 184–85

Fort Francis E. Warren, Wyo., 245

Fort Fred Steele, Wyo., 88

Fort George Wright, Wash., 213, 216

Fort Grant, Ariz., 181

Fort Hall, Idaho, 25

Fort Hancock, N.J., barracks, *235*, *236*

Fort Harstuff, Neb., 145, 261n.47

Fort John (trading post), *14*, 25, *26*

Fort Kearny, Neb., 17, 25, 26, 73

Fort Keogh, Mont., 153

Fort Laramie, *14*, *31*, *32*, *40*, *146*, *189*; administration building, 99, *100*, 145; animals, 36; architectural variety, 133–34; bakery, 145, 153; barracks, 47–49, *50*, *51*, 65, 98, 99, 45, 146, 184, 185, 195; batteries, 33; Bedlam, 45–47, *46*; blockhouse, 31, 33–34; chapel, 99; civilian employees, 7, 64, 113; civilian employees' quarters, 45, 49–50, *53*; closing, 133, 203, 245; construction practice, 216; corral, 33–34; and emigrants, 14, 17; fortifications, 30–34; guard house, 31, 45, 50, *54*, 145, 152; hospital, *105*, 118, 145, 155, *162*, *165*, 165–66; hospital garden, 102–103; latrine, 186–87; laundresses' quarters, 49, *52*, 145; layout, 36–37; library, 99; location of, *15*, 23–27; noncommissioned officers' quarters, 137, 145; officer's quarters, *37*, 45–47, *46*, *48*, 51, 74, 75, 98, *103*, *106*, 113, *135*, 145, 149–50, *153*, *154*, *170*, 177–78, 193; parade ground, 37, 101; post-army, 247–49; post garden, 102; post trader, 114–17; post trader's house, *55*, 51–52, 115; post trader's store, 45, 52, 115–17, *116*; powder magazine, 45; purpose, 9, 14–15, 27, 98; redoubt, 33–34; sawmill, 45, 67, *68*, 68; shingle mill, 69; shops, 45, 51; soldiers, 108; storehouses, 145; supply system, 67, 91, 140, 142; teamsters' quarters, 33–34; trading post, 23–25, 45, 73; vicinity, 89; walls, 31; water distribution, 187, 192; yards, 104–106

Fort Laramie National Monument, 249

Fort Larned, Kans., 20

Fort Leavenworth, Kans., 198, 265n.6

Fort Lincoln, N. Dak., 213, 219

Fort Logan, Colo., 212

Fort Lyon, Colo., 256n.1
Fort Mackenzie, Wyo., 213, 215, 219
Fort McKinney, Wyo., 69, 209, 216
Fort McPherson, Neb., 175–76
Fort Meade, S. Dak., 3, *6*, *104*, 153, 177
Fort Missoula, Mont., 265
Fort Monroe, Va., 225
Fort Myer, Va., 219
Fort Niobrara, Neb., 262n.66
Fort Omaha, Neb., 213
Fort Phil Kearny, Wyo., 8, 20, 23, 29, *30*, 45, 62, 68, 69, 71–72, *72*, 101–102, 111
Fort Pierre, S. Dak., 141
Fort Reno, Wyo., 29, 258n.34
Fort Reynolds, Colo., 62
Fort Riley, Kans., 198
Fort Robinson, Neb., 69, 146–47, 208, 213
Fort Russell Progressive Club, 238
Forts: design, 5, 9–10, 16; fortifications, 29–35; function, 17; landscaping, 99–107; layout, 36–38; location, 18–20, 21, 23–29; parade ground, 36; permanent, 138–39, 140; sanitation, 186–96, 240–41; temporary, 22, 138–39
Fort Sanders, Wyo., 72
Fort Sedgwick, Colo., 19
Fort Shaw, Mont., *183*
Fort Sheridan, Ill., 198
Fort Snelling, Minn., 241
Fort Stevens, Ore., 212
Fort Thornburgh, Utah, 73, 131
Fort Totten, N. Dak., 140
Fort Union, N. Mex., 74, 256n.1
Fort Vasquez, Colo. (trading post), 73
Fort Washakie, Wyo., 208
Fort William (trading post), 24, *25*
Fort William Henry Harrison, Mont., 213
Fowler, Orson S., 144–45
Francis E. Warren Air Force Base, 245, *246*, *247*
Freeman, Mary E. Wilkins, 92
Funston, Frederick, 240
Furnishings, 65, 175–82, *177*, *178*, *180*, *182*, *183*, *237*, 240

Gillmore, Quincy, 145
Glassie, Henry, 126
Grand Rapids Furniture Company, 240
Grant, Ulysses S., 18, 19, 20, 64
Grattan, John, 17
Grattan Massacre, 17
Great Sioux Reservation, 20, 27
Grimes, Edward B., 29

Grodavent, F. J., 212
Grummond, George, 258n.45

Hall, William P., 149
Hancock, Winfield S., 44
Harding, Warren G., 205
Harney, William S., 17
Harrison, Benjamin, 209
Havard, Valery, 66
Hayes, Rutherford B., 116
Hazen, William B., 29
Hobbs, L. (mechanic), 144–45
Hoffman, William, 68, 73
Holabird, Samuel B., 112, 137, 139, 169, 174, 194
House at 414 E. 22nd St., Cheyenne, 127, *130*
Howard, Oliver O., 140
Howe, Henry S., 66
Hudson Bay Company, 29
Hughes, William B., 190
Humphrey, Charles F., 125, 169, 193, 195, 213–15
Hunton, John, 247

Indians, 6, 9, 14–21, 27, 29, 33–35, 36, 49, 82–88, 91, 117–18, 203, 208. *See also* individual tribes and names
Iron Bull, 117

Jeffrey, Julie Roy, 111
Jerrett, George H., 132, 168
Jesup, Thomas S., 21, 23
Jewett, Sarah Orne, 92
Johns Hopkins Hospital, 158
Johnson & Davis, 219
Johnson County War, 208–209
Johnston, Albert S., 35
Julien, J. P., 127

Kansas Pacific Railroad, 82
Keefe, Moses P., 127, 166, 216, 219, 222
King, Charles, 46, 120, 129
King Bridge & Manufacturing Company, 142

LaFollette, Robert, 264n.2
Lincoln General Hospital, 257n.26
Lord, James, 125

Mackay, Aeneas, 26
Mahan, Dennis Hart, 63
Manhattan Brass Company, 197

Marcy, Randolph B., 62, 136, 148
Marcy Board (Board on Revision of Army Regulations), 135, 148
Maynadier, Henry E., 62
McCrary, George W., 83
McKim, Mead & White, 214
McKinley, William, 206
McParlin, Thomas A., 147
Meeker, Ezra, 247
Meigs, Montgomery C., 18–20, 43, 58–59, 64, 72, 74–75, 106, 115, 120, 121, 135–36, 148–52, 184
Meinig, D. W., 79
Merritt, Carrie, 172
Merritt, Wesley, 172
Miles, Nelson, 206
Military Service Institution, 265n.6
Millard, Bailey, 241
Miller, Leslie A., 249
Minuteman missiles, 245
Mizner, Henry, 193
Model plans, 41–44, 134, 147–53; bake house, 153, *161*; barracks, *42, 149, 150, 151*, 184, 185; bath house, 195; guard house, 152, *158*; hospital, 162–67, *163*; officers' quarters, *42, 43*, 149, *152, 155*, 169–70. *See also* Standard plans
Mormons, 9, 14, 17, 27, 28, 35, 53, 132, 248
Mountain Man Rendezvous, 249
Murdock, Charles, 226
Murphy, S. J., 191–92
Myers, William, 44, 141

Nagel-Warren House, Cheyenne, 94
New York Hospital, 166
New York Public Library, 158

Ogle, Alexander, 164
Oregon-California Trail, 9, 17, 20, 25, 27–28, 32, 86, 98, 247–48, 249
O'Reilly, Robert M., 165

Pacific Railroad Act (1862), 86
Page, Frank A., 135
Pardee, Julius H., 121
Patten, William S., 227
Peace Commission, 20, 27
Peacekeeper missiles, 245
People at fort: children, 113; civilian employees, 113; emigrants, 14–15, 17, 21, 27, 117; enlisted men's wives, 109, 111; laundresses, 109–11, *110*; officers, 108; officers' wives, 111–13, 172–73; post trader, 114–17; servants, 178–79, 217, 220, 223–24, 227; settlers, 82, 85; soldiers, 108, 241; soldiers, African-American, 108, 241
Perry, Alexander, 145
Pine Ridge Agency, S. Dak., 208
Poinsett Plan, 16
Pope, John, 83, 151
Porter, Jeremiah, 99
Post, George B., 166

Quartermaster Replacement Training Center, 245

Railroad, 67, 82, 85, 86, 98, 139, 141, 172, 207, 208
Ramsay, Alexander, 83
Red Cloud, 20, 27
Reed, Walter, 147
Regan, James, 190, 196
Reps, John, 80
Robertson, J. M., 148
Robinson, Henry E., 104
Rodgers, John F., 181
Rogers, W. W., 143–45
Roosevelt, Franklin D., 249
Root, Elihu, 206–207, 211, 214
Rusling, James F., 28
Russell, David A., 86
Russell, Majors & Waddell, 17
Russell Sanders Construction Company, 216
Rustic House, 116

Sand Creek, Colo., 18, 33
Sanderson, Winslow F., 26
Schell, Henry S., 155
Schofield, John M., 209, 265n.6
Scott, Winfield, 21
Shellenburg, J. M., 240
Sheltering fund, 19
Sheridan, Philip, 22, 83, 145, 188
Sherman, William T., 6, 18–20, 21, 22, 34, 63, 82, 85, 86, 92, 98, 111, 112, 124, 131, 138, 140, 148, 173, 178, 203, 205
Shoshoni Indians, 18
Sidney Barracks, Neb., 143–45
Simonin, Louis, 46, 52
Simpson, James F., 195
Sioux Indians, 17, 20, 29, 30, 83, 91, 208
Sitting Bull, 83
Smart, Charles, 55, 185

Smith, Addison, 248
Smith, Fred A., 224
Snyder, Simon, 62
Spanish-American War, 22, 206
Standard plans, 170–71, 211–36, 242–43; gymnasium, *239*; officers' quarters, *218, 222, 224, 228, 229*. *See also* Model plans
Stanton, William S., 142
Steele, W. R., 142
Stevenson, John D., 86, 91, 94
Stivers, Charles P., 131, 190
Stowe, Harriet Beecher, 92
Strategic Air Command, 245
Sublette, Andrew, 73
Sublette & Campbell, 24
Summerhayes, Martha, 119–20, 176, 178–79
Swan, Alexander, 91

Taft, William Howard, 207, 210
Three Mile Ranch, 145
Totten, Joseph G., 32
Treaty of Fort Laramie (1851), 17, 27; (1868), 20, 82
Turnley, Parmenas T., 141

Union Pacific Railroad, 9, 82, 85, 86, 88, 89, 91–92, 140, 190, 198, 204, 209
Upton, Emory, 85
U.S. Army: role in West, 6–7, 9, 14–23, 33–35, 81–88, 89, 203; structure of, 22, 204–207
U.S. Capitol, 19, 10, 21–22
U.S. Congress, 22, 25, 75, 84, 138–39, 142, 204–207, 215, 241
U.S. National Park Service, 249
Ute Indians, 18, 88, 131

Van Brunt & Howe, 214
Van Vliet, Stewart, 102
Van Voast, James, 33
Vasquez, Louis, 27

Wabash College, 62
Ward, Seth E., 52, 114, 115
Waring, George, Jr., 81
Warren, Francis E., 204–208, 214, 215, 236–37, 240, 244
Warrens, Charles H., 68, 145, 149
Weigley, Russell, 85
West Point, 19, 41, 62–63, 85, 168–69
Wheaton, Francis B., 214, 221
Wherry housing, 245, *247*
Wind River Agency, Wyo., 208
Wister, Owen, 198
Wood, Joseph, 80
Wood, Leonard, 207
Woodbury, Daniel, 26, 30–32, 37, 45, 73, 140
Wounded Knee, Battle of, 83, 203, 208
Wright, Luke, 215
Wyoming Archives and Museums Historical Department, 248
Wyoming Governor's Mansion, 226–27, *231*
Wyoming Historical Landmark Commission, 248–49
Wyoming State Capitol, 92, 216
Wyoming State Library, Archives, and Historical Board, 248
Wyoming Stock Growers Association, 91, 208–209

Yorktown Centennial, 192
Young, Brigham, 27